Thomas Howard

On the Loss of the Teeth and Loose Teeth

And on the Best Means of Restoring them

Thomas Howard

On the Loss of the Teeth and Loose Teeth
And on the Best Means of Restoring them

ISBN/EAN: 9783337429393

Printed in Europe, USA, Canada, Australia, Japan

Cover: Foto ©berggeist007 / pixelio.de

More available books at **www.hansebooks.com**

PARIS

IN OLD AND PRESENT TIMES

*WITH ESPECIAL REFERENCE TO CHANGES IN ITS
ARCHITECTURE AND TOPOGRAPHY*

BY

PHILIP GILBERT HAMERTON
Officier d'Académie

WITH MANY ILLUSTRATIONS

BOSTON
ROBERTS BROTHERS
1885

PREFACE.

IT is probable that there is not another city in the whole world that has undergone so many and such great changes as the capital of France. Those of us who have been familiar with Paris since the accession of Louis Napoleon have been eye-witnesses of the last of these, which consisted chiefly in improving the means of communication by opening wide new streets, and in erecting vast numbers of houses of a new type. From the sanitary point of view the change was most desirable and circulation was made incomparably easier; from the artistic point of view there was a balance of loss and gain, as the old streets were not always, or often, worth preserving, while the new ones have always some pretension, at least to taste and elegance, and many new buildings are really good examples of modern intelligence and art. But there is a certain point of view from which this reconstruction of an ancient city was entirely to be regretted. Archaeologists deplored the effacement of a thousand landmarks, and if it had not been for their patient labors in preserving memorials of the former city on paper, the topography of it would have been as completely iv *Preface.*

effaced from the recollection of mankind as it is from the actual site. Were it not for the existence of a very few old buildings such as Notre Dame, the Sainte Chapelle, the Hôtel de Cluny, and one or two other remnants of past architectural glories, Paris might seem to date from the age of Louis XIV.; and even the remaining works of the great king are not sufficiently numerous to give an aspect to the city, which seems as new as Boston or New York,—I had almost written, as Chicago. While Avignon and Aiguesmortes preserve their ancient walls, the *enceinte* of Paris has been repeatedly demolished, carried farther out, and reconstructed on new principles of fortification. While the palace of the Popes still rears its colossal mass on its rocky height near the Rhone, and withstands, unshaken, the unequalled violence of the mistral that sweeps down upon Avignon, the palace of the mediaeval kings has almost entirely disappeared from the island in the Seine, and the old Castle of the Louvre is represented by an outline in white stone traced in the pavement of a quadrangle. Of the wall of Philippe-Auguste the very last tower has long since disappeared, and the grim fortress of the Bastille has utterly vanished from its site, known to modern Parisians as a stopping-place for omnibuses. Nor has the more modern palace of the Tuileries escaped a similar annihilation. The last stone of it was carted away not long since, and our best record of its ruin is a little study or picture by Meissonier. Every year it becomes less and less profitable to visit Paris in ignorance of its past history; and therefore it has seemed to me that such an account of the city as I should care to write must include constant reference to what has been, as well as a sufficiently clear description of what is. This has not been done before in our language, and would not have been possible now if the admirable labors of many French archaeologists had not supplied the materials. I need not add that whenever anything could be verified by personal observation, I have taken the trouble to see things for myself. Paris has been very familiar to me for nearly thirty years; but in spite of this long intimacy with the place, I went to stay there again with a view especially to the present work.

I may add that, although I have written little hitherto about architecture, it has always been a favorite study of mine, and I have neglected no opportunity of increasing such knowledge of it as a layman may possess. The facts about the history and construction of edifices given in the present volume may, I believe, always be relied upon; as for mere opinions, I give them for what they may be worth. The best way is for a critic to say quite candidly what he thinks, but not to set up any claim to authority.

CONTENTS.

		PAGE
I.	INTRODUCTION	1
II.	LUTETIA	16
III.	A VOYAGE ROUND THE ISLAND	34
IV.	NOTRE DAME AND THE SAINTE CHAPELLE	59
V.	THE TUILERIES AND THE LUXEMBOURG	82
VI.	THE LOUVRE	104
VII.	THE HÔTEL DE VILLE	125
VIII.	THE PANTHEON, THE INVALIDES, AND THE MADELEINE	139
IX.	ST. EUSTACHE, ST. ETIENNE DU MONT, AND ST. SULPICE	159
X.	PARKS AND GARDENS	174
XI.	MODERN PARISIAN ARCHITECTURE	197
XII.	THE STREETS	219

LIST OF ILLUSTRATIONS.

	PAGE
TRANSEPT OF NOTRE DAME	4
OLD HOUSE WITH TOURELLE	9
THE HÔTEL DE CLUNY	12
THE OLD MAISON DIEU, AND NORTH TRANSEPT OF NOTRE DAME	14
THE FRIGIDARIUM OF THE ROMAN BATHS, CALLED LES THERMES	20
THE GRAND CHÂTELET	28
THE TOUR DE NESLE. FROM THE ETCHING BY CALLOT	30
THE LOUVRE OF PHILIPPE-AUGUSTE	32
GARDEN EAST OF NOTRE DAME	40
PONT NOTRE DAME, 18TH CENTURY	44
THE PUMP NEAR THE PONT NOTRE DAME, 1861	46
THE PONT NEUF IN 1845	50
THE MORGUE IN 1840	52
THE LITTLE CHÂTELET, TAKEN FROM THE PETIT PONT IN 1780	54
THE ARCHBISHOP'S PALACE IN 1650. FROM AN ETCHING BY ISRAEL SYLVESTRE	56
ANGLERS ON THE QUAYS	58
TYMPANUM OF THE PORTE STE. ANNE	64
PIER AND ONE OF THE DOORS OF THE PORTE STE. ANNE	66
LES TRIBUNES	68
THE "POURTOUR"	70

List of Illustrations.

	PAGE
ROYAL THANKSGIVING IN NOTRE DAME, 1782	74
THE OLD COURT OF ACCOUNTS AND THE SAINTE CHAPELLE	78
SAINT LOUIS IN THE SAINTE CHAPELLE	80
THE TUILERIES IN 1837	96
THE LUXEMBOURG AS IT WAS BUILT	100
THE LOUVRE IN ITS TRANSITION STATE FROM GOTHIC TO RENAISSANCE	104
THE LOUVRE, FROM THE SEINE. FROM A DRAWING BY H. TOUSSAINT	106
DETAILS BY PIERRE LESCOT IN THE QUADRANGLE	107
THE CLASSICAL PAVILION AND THE OLD EASTERN TOWER	110
THE INTERIOR OF THE QUADRANGLE. FROM A DRAWING BY H. TOUSSAINT	114
QUADRANGLE OF THE LOUVRE, WITH THE STATUE OF FRANCIS I., PLACED THERE IN 1855, AND SINCE REMOVED	118
THE COLONNADE. FROM A DRAWING BY H. TOUSSAINT	120
PERRAULT'S COLONNADE. INTERIOR VIEW	122
AN OLD ROOM IN THE LOUVRE	124
FRONT OF THE HÔTEL DE VILLE IN THE TIME OF LOUIS XIII.	128
THE HÔTEL DE VILLE IN 1583. FROM A DRAWING BY JACQUES CELLIER	130
THE GREAT BALL-ROOM	136
THE PANTHEON	142
THE PANTHEON FROM THE GARDENS OF THE LUXEMBOURG	146
THE INVALIDES	152
THE MADELEINE	155
THE CHURCH OF ST. EUSTACHE	160
CHURCH OF ST. ETIENNE DU MONT. FROM SKETCH BY A. BRUNET-DEBAINES	162
INTERIOR OF ST. ETIENNE DU MONT	164
WEST FRONT OF ST. ETIENNE DU MONT	168
THE CHURCH OF ST. SULPICE	170
GRANDE ALLÉE DES TUILERIES	181

List of Illustrations.

	PAGE
LAC DES BUTTES CHAUMONT	183
LE TROCADERO	184
AVENUE DES CHAMPS ELYSÉES	186
AU JARDIN DU LUXEMBOURG	188
LAC DU BOIS DE BOULOGNE	190
LA NAUMACHIE, — PARC DE MONCEAU	192
DOORWAY OF A MODERN HOUSE	204
THE OPERA. SIDE VIEW	206
THE OPERA. THE PRINCIPAL FRONT	208
INTERIOR OF THE CHURCH OF ST. AUGUSTINE	210
THE CHURCH OF ST. AUGUSTINE	212
INTERIOR OF THE CHURCH OF LA TRINITÉ	214
THE CHURCH OF LA TRINITÉ	216
BOULEVARD ST. GERMAIN	222
AVENUE FRIEDLAND	228
HÔTEL DE SENS	230
THE MAIRIE AND ST. GERMAIN L'AUXERROIS	234
RUE DES CHIFFONNIERS, PARIS. DRAWN BY LEON LHERMITTE	236

PARIS

IN OLD AND PRESENT TIMES.

I.

INTRODUCTION.

NATIONALITY affects our estimates of everything, but most especially does it affect our estimate of great cities. There is no city in the world that does not stand in some peculiar relation to our own nationality; and even those cities that seem quite outside of it are still seen through it, as through an atmosphere colored by our national prejudices or obscured by our national varieties of ignorance.

Again, not only does nationality affect our estimates, but our own individual idiosyncrasy affects them to a degree which unthinking persons never even suspect. We come to every city with our own peculiar constitution, which no amount of education can ever alter fundamentally; and we test everything in the place by its relation to our own mental and even physical needs. We may try to be impartial, to get at some sort of abstract truth that has nothing to do with ourselves;

but it is not of any real use. There is a certain relation between human beings and places which determines, in a wonderfully short time, to what degree we are capable of making ourselves at home in them, — how much of each place belongs to us by reason of the obscure natural affinities.

Before entering upon this great subject, Paris, I think it will not be a waste of space, or a useless employment of the reader's time, if I show in what way our estimate of that city is likely to be affected by our national and our personal peculiarities.

First, as to nationality. Englishmen admire Paris; they speak of it as a beautiful city, even a delightful city; but there is one point on which a Frenchman's estimate of Paris usually differs from that of an Englishman. I am not alluding to the Frenchman's patriotic affection for the place; that, of course, an Englishman cannot have, and can only realize by the help of powerful sympathies and a lively imagination. I am alluding to a difference in the impression made by the place itself on the mind of a French and an English visitor. The Englishman thinks that Paris is pretty; the Frenchman thinks that it is sublime. The Englishman admits that it is an important city, though only of moderate dimensions; the Frenchman believes it to be an immensity, and uses such words as "huge" and "gigantic" with reference to it, as we do with reference to London. Victor Hugo compares Paris with the ocean, and affirms that the transition from one to the other does not in any way exalt one's ideas of the infinite. "*Aucun milieu n'est plus vaste,*" he says, very willingly leaving the much larger British capital out of consideration. For him Paris is everywhere, like the air, because it is ever present in his thoughts. " *On regarde la mer, et on voit Paris.*"

We Englishmen, always remembering London, and more or less consciously referring every city to that, are very liable to a certain form of positive error with regard to Paris, against which, if we care for truth, it is well to put ourselves on our guard. Most things in Paris seem to us on rather a small scale. The river seems but a little river, as we so easily forget its length and the distance of Paris from the sea; and most of the buildings that Englishmen care to visit are near enough to their usual haunts to produce the impression that the town itself is small. The Louvre, the Luxembourg, Notre Dame, the Madeleine, the Opera, and the *Palais de l'Industrie*, are included within that conveniently central space which to the Englishman is Paris. Even the very elegance of the place is against it, insomuch as it produces an impression of slightness. A great deal of very substantial building has been done in Paris at all times, and especially since the accession of Napoleon III.; yet how little this substantial quality of Parisian building is appreciated by the ordinary English visitor! I remember making some remark to an Englishman on the good fortune of the Parisians in possessing such excellent stone, and on their liberal use of it, and on its happy adaptability to the purpose of the carver. The only answer I got was a laugh at my own simplicity.

"That white stuff is not stone at all; it's only stucco!" This observer had seen hundreds of carvers chiselling that stone, yet he went back to London complacently believing that all its ornaments were cast. Here you have a striking example of patriotic error, — the stone of a foreign city believed to be stucco because stucco is a flimsy material, and because it was not agreeable to recognize in foreign work the qualities of soundness and truth. Even in this mistake may be traced the pre-disposing influence of London. Stucco has been used in very large quantities in London; and the stone employed there in public buildings, though of various kinds, is never of the kind most extensively employed in Paris.

It is unnecessary to dwell any longer upon what Mr. Herbert Spencer would call the "patriotic bias." French people bring the same bias with them into England, and write accounts of London with astounding inaccuracy. In one of the most recent of these there occurred a description of the House of Lords, giving no idea whatever of its architecture, and stating that it was not bigger than an ordinary council-room in a provincial *mairie*.[1] Many things in London are as heartily despised by intelligent Englishmen as they can possibly be by foreigners, but the foreigner shows his own patriotic bias by dwelling upon them, and by slighting allu-

[1] I am inclined to think that the Frenchman's notions of size had been upset by passing through Westminster Hall; but the patriotic bias in his account of the Houses of Parliament was shown by his omission of architectural appreciation, and by his extreme readiness to describe what he supposed to be eccentricities or defects.

TRANSEPT OF NOTRE DAME.

sions to what is really good and noble in London, — for example, when he passes by St. Paul's as a feeble imitation of St. Peter's at Rome, or speaks of the Law Courts as a medley of Gothic details, without doing justice to the originality of either Wren or Street. A French critic is usually so horrified by London smoke and by the ugliness of our ordinary houses, that he becomes incapable of perceiving beauty even where it really exists, and confounds all things together in undiscriminating, unsparing condemnation.

From these influences of nationality I do not hope to be wholly free, though at the same time I am neither conscious of any patriotic bias against the capital of France, nor of any anti-patriotic bias in its favor. I have been very familiarly acquainted with Paris for twenty-seven years, and know both its beauties and its defects. The only strong national prejudice against it which I still retain is a rooted prejudice in favor of the old English system of living in separate houses as against the French system of living on flats. It may seem at first sight that this has very little to do with the artistic aspects of Paris, which will be the subject of the present series of papers; but, in truth, the connection between them is very close. The magnificence of modern Parisian streets is almost entirely due to the flat system; the apparent meanness of English towns is due to our separate houses. I am quite aware of this; and I know at the same time that where land is expensive, as it must be in every great city, the flat system is the one which allows the widest and most spacious streets, and gives the most air and sunshine to the inhabitants. Still, while admitting the convenience of the arrangement, its reasonableness, and the architectural grandeur of the combinations that result from it, I am Englishman enough to prefer, in my heart of hearts, a quiet English house with a ground-floor and one upper storey, or two at the very utmost, to the most imposing and pretentious pile of towering *appartements* that the skill of a French architect ever devised or the wealth of an American colony ever rented. I revisited the north of England towards the close of 1882, and remember thinking, at Burnley, that one of the clean little houses that are now built there for workpeople, each with its own independent entrance and ready access to the street, would be pleasanter to live in than an expensive *appartement au quatrième* on one of the finest boulevards of Paris. This no doubt is an English prejudice; but one cannot denationalize oneself altogether.

With regard to personal as distinct from national prejudices, the only important one that I am conscious of is a strong dislike to such extension of size in towns as that which makes them rather regions covered with houses than creations complete in themselves. A city of small size (what a Londoner would call insignificant), well situated in beautiful scenery, with ready access to the country from all its streets, and itself so constructed that its principal edifices compose happily with the landscape, and adorn it, — this is my ideal of a town; an ideal not so far from a possible reality, but that there are actually some existing little cities in France and

Italy that respond to it. The complete opposite of this ideal is London, which is not a town, but a spreading and gathering of population, like irregular fungoid growths joining together by their edges till a great space is ultimately covered by them, while there seems to be no reason why they should not spread indefinitely on every side. There is nothing, on the outskirts of London, of that pretty, sudden contrast between town and country which gives such charm to both when the real green country, with its refreshment of rural peace, comes close up to the gray walls of the city, and shades them with its trees and adorns them with its flowers; when the citizen can be at his business in the heart of the city at sunset and in the quiet fields before the gold has faded from the evening sky. That time is past for Paris as for London; but some names of places still remain to recall rural associations. *St. Germain-des-Prés*, now close to a noisy boulevard, was once an abbey-church among meadows; *Notre Dame des Champs* was really Our Lady of the Fields; and the *Rue Neuve des Petits Champs*, a new street in little fields. Primroses may once have been found in the *Impasse des Primevères*, and vines in the *Impasse des Vignes*. The country came close up to the smaller Paris of the Middle Ages, and round about it there were fortresses, monasteries, and villages, islanded in a sea of pasturage, corn, and vines. Wall after wall was found to be too narrow a boundary, till M. Thiers built the present fortifications, which the municipal council, with the consent of the military authorities, are already disposed to demolish, except the detached forts. This continual expansion of Paris beyond its boundaries, this continual invasion of the surrounding country, has given to the city that ill-defined zone of cheap and hasty construction which surrounds every growing town. There is no longer a complete Paris, that can be easily seen at once. Giffard's captive balloon gave the means of seeing the present Paris, which presented the appearance of a vast basin covered with houses that died away into the surrounding country, and were divided by a many-bridged river; but the balloon was wrecked by a tempest, and now it is only the adventurous free aeronauts who, as they drift about in the upper air at the wind's will, can see the great city of the Seine.

It is a convenience to divide history into epochs, which we select to mark the accomplishment of great changes; but this habit of arbitrary division conveys in one way a false impression to the mind. The changes seem complete when we speak inaccurately and generally; but if we look carefully and strictly into the matter we shall find that every age has left its peculiar work unfinished, and has left it to be continued by the next age, which, in its own turn, has begun something else, and left that to be carried on by its successor. There appears to be no such thing as finality in the history of a great city; and, indeed, we may conclude from what has been actually done by past generations, that there is no incentive to important public works so powerful as the continual appeal of half-executed projects. The stones of many a building call as loudly as if they could

OLD HOUSE WITH TOURELLE.

really speak; they call not only for care in their preservation, but for additions to make them look less forlorn. Sometimes too much is done; mistakes are committed that need correction, and new mistakes are made in trying to rectify old ones, or a certain thing is built that would have been complete in itself if it could only have been let alone; but it was not big enough for subsequent practical needs, and so additions were made which destroyed its proportion, as if the wings of an eagle were fastened to a sparrow-hawk. Only a very few buildings, either in Paris or any other modern city, have possessed the virtue of unity.

We ourselves have witnessed one of the most complete transformations of Paris. We have seen the Paris of Louis-Philippe transformed into that of Napoleon III.; but even this, the greatest change ever operated in so short a time, had been prepared for, as I shall demonstrate when we reach that portion of our subject, by architectural tendencies and practical necessities which had been seen and felt much earlier. A much more absolute distinction exists between Gothic Paris and the Paris of the Renaissance. There, indeed, was a radical change, right and necessary as preparing the way for modern life, but at the same time exceedingly destructive, and not by any means generally favorable to grace or beauty in its beginnings. It would be easy to describe the Paris of Louis XI. in very eloquent language, by the simple process of bringing every beauty into brilliant relief and hiding every defect, and it would be not less easy to make it appear that the Paris of Louis XIV. was a heavy and expensive mistake; but we shall have no controversial purpose to answer in this book. The course of events by which a beautiful and convenient modern city has replaced a picturesque mediaeval one, is full of interest to the student, but need not awaken in him any very deep sentiment of regret, unless it be for this or that particular building which he knows to have once stood where omnibuses are now running on the Boulevard, or cafés display their vulgar luxury close by. This is the way in which our loss is most effectually brought home to us. There is the Hôtel de Cluny, for example, which has been preserved almost by miracle down to the present time, and is now made as safe for the future, by legislative protection, as any human work well can be. Go through that admirable dwelling, so charming in its variety, without any violation of harmony, so unostentatious and yet so beautiful, so well adapted to the needs of honorable and peaceful human life, and then calculate how many furlongs of monotonous modern houses in the Rue de Rivoli might possibly be accepted as an equivalent for it. The Hôtel de Cluny is the best of the old houses now remaining, almost the only important one that is still anything better than a fragment; but historical students go from site to site, where the best of the old dwellings used to be, and then, finding nothing equivalent in their places, they lament what seems to them a blank, uncompensated loss. The loss is seldom compensated for on the spot, or in anything of the same kind; but there is a broader and more general compensation in the grandeur of the modern city. If Paris had been treated somewhat tenderly, as Bourges has been, if the mediaeval houses had been generally preserved, and consequently the mediaeval streets, the houses keeping their external appearance and being adapted to modern requirements by internal alterations only, then indeed the city would have been a pleasant place for the investigations of the artist and the archaeologist; but communication would have been so difficult that the life-blood of a great and populous modern city could never have circulated through such narrow and frequently constricted arteries. Nor has the destruction been quite absolutely complete. Notre Dame and the Sainte Chapelle have been preserved at least as well as Westminster Abbey and the Temple Church, while the tower of St. Jacques is left standing, when the church itself is gone. The less important remains of the Middle Ages, a small house or a *tourelle* here and there, were rapidly disappearing in Méryon's time, and with few exceptions have vanished utterly since.

In Victor Hugo's "Notre Dame de Paris," written in 1830, after a long and brilliant description of Paris in the Middle Ages, there comes a prediction of evil omen which has happily not been realized. "Our fathers," he says, "had a Paris of stone; our sons will have a Paris of plaster."

"The Paris of the present day (1830) has no general character. It is a collection of specimens of different ages, and the finest have disappeared. The capital increases only in houses — and what houses! At this rate there will be a new Paris every fifty years. And then the historical significance of its architecture is effaced daily. Buildings of importance become rarer and rarer, and it seems as if we could see them gradually sinking — drowned in the flood of houses. Our fathers had a Paris of stone; our sons will have a Paris of plaster."

THE HÔTEL DE CLUNY.

Introduction.

This city of plaster might have filled the whole space within the fortifications to-day if the railways had not brought stone so easily from a distance; but by a happy coincidence the colossal building enterprises of Napoleon III. were not undertaken before the principal lines of railway had been constructed, and by their means, not stone only, but vast quantities of wood and other materials were brought readily to hand. At the same time the feeling, which an enemy calls vanity and a friend self-respect, led the sovereign and the municipal authorities of that time to desire that the new Paris should be a credit to them, — one of the principal glories of what was intended to be a very brilliant reign. The consequence has been the reverse of what Victor Hugo feared. The Paris of plaster was the capital of Charles X. and of Louis-Philippe. Miles and miles of new streets were driven through dense clusters of houses so slight and poor in construction that they only kept themselves from falling by leaning against each other, while they did not possess the slightest architectural merit. In the new streets the houses were built of stone, and the work was done to endure. Of this new stone Paris we shall have much to say in this volume. The greatest fault of it is a certain monotony; but this was especially the fault of the first attempts in the new style.

During the later years of Napoleon III., and since his time, there has been more variety in Parisian street architecture, though it is true that the variety is often rather in the invention of detail than in the conception of edifices. There are immense quantities of good ornamental sculpture, by no means slavish in the copying of set types, but full of delicate fancy, and really of our own time, though deriving its origin from the best French Renaissance. In a word, there is really a living street architecture in Paris in which clever architects employ ingenious artists and highly trained craftsmen to work upon the best materials. What remains true in Victor Hugo's criticism is, that the great height of these modern houses, and their enormous quantity, make public buildings seem as if they were drowned among them. All the churches in Paris, not excepting Notre Dame, have been diminished by gigantic modern house-building; just as a great injury has been done to the National Gallery, in London, notwithstanding its very favorable site, by the neighborhood of the Grand Hotel. We remember the time when the Nelson Column used to appear unnecessarily high, but it is not an inch too high at present; and we all know what a deplorable effect has been produced upon the towers of Westminster Abbey by the tall new houses in their neighborhood. So the greater decorative enrichments of modern buildings have often made an older edifice look poor, as Westminster Hall was externally annihilated by the panelled walls of the new palace, and the old Tuileries made to look poverty-stricken beside the massive ornaments of the new *Pavillon de Flore*. Hence it is a most dangerous time for the public buildings in any city when the people are beginning to take a delight in lofty houses and palatial hotels. Nor is this danger

THE OLD MAISON DIEU, AND NORTH TRANSEPT OF NOTRE DAME.

confined to cities only; an old building of moderate dimensions, even in the country, may be reduced to nothing by a large new one erected near enough to it for comparison. They tell me that a great hotel has been set up very near Kilchurn Castle. The only tolerable thing near the moderately sized castles of the Highlands is a lowly thatched cottage, with green moss on its roof, and blue peat-reek rising through a hole in it.

II.

LUTETIA.

IT is curious that the sites of the most important cities in the old world should generally have been determined by the choice made by a barbarous tribe thousands of years ago, with a view to its own security, and that this choice made by barbarians should have settled the matter so irrevocably that succeeding generations have had to do the best they could with the same position, well chosen for the needs of its first occupants, but often ill chosen for the latest. The selection of Paris as the site of the future capital of France depended on the practical wisdom of some prehistoric savages, who found that islands in the river were the safest places to be had in that part of the country. There was one large island, and a few smaller ones, in the midst of the tract of country now occupied by Paris, and there is evidence that some prehistoric tribe used these islands for a protected dwelling-place. After them came the Gauls, with a far higher degree of civilization and a rather advanced military art, especially in defensive arrangements. The Gaulish oppidum was not what we understand by a city, even when the city is fortified; it was simply a place of refuge, in some situation naturally difficult of access, either from steepness, as in hilly countries, or from bogs and water in more level ones. The Gauls preferred a steep hill to anything else as the site of one of their great forts; but where they had not a hill high enough and steep enough for their purpose, they were glad of a piece of solid ground in the middle of a marsh, or an island in a river. The island on which Notre Dame is now situated appears to have answered their purpose, and for long afterwards its defensive value was of some consequence; but I need hardly observe that when Paris was besieged by the Germans in 1870, it did not signify in the least whether the central part of the city was on an island or not. Paris has so immensely outgrown its first insular beginning, that its present military defences are a ring of forts far away out in the country on all sides. I am rather inclined to believe that in this extension we may see a prototype of Great Britain, scarcely to be considered an island since her Colonial Empire has become so vast as to give her frontiers inside three continents.

The numbers of bridges in Paris make the islands as much a part of the town as any other part, and indeed we are hardly sensible that they are islands at all. But not only was the Gaulish oppidum insular, the Gallo-Roman city of Lutetia was so too; and there is every reason to believe that it presented rather a beautiful appearance as seen from the surrounding country. In Hoffbauer's valuable work on "Paris à travers les Ages," to which I am under great obligations for archaeological details not readily accessible elsewhere, there is a careful drawing of Lutetia as it must have appeared from the aqueduct of Arcueil, with Montmartre, then the Hill of Mars, in the distance. The first impression one receives is that, compared with mediaeval Paris, Lutetia must have had a strangely modern look; but the fact is, that since the Renaissance we have got so thoroughly used to classic forms that we are really at home in them, and it is positively more natural for us to build (with certain modifications) like the ancient Romans than like our own mediaeval ancestors. The aqueduct of Arcueil in M. Hoffbauer's drawing reminds one of a suburban railway viaduct; the Roman villas among the trees in the valley are in outward appearance not very unlike many French and Italian houses of the present day; and if Lutetia on her island has an aspect rather unsatisfying to modern eyes, it is more because there are neither domes nor spires nor any lofty towers, than because the edifices themselves are contrary to our taste.

The Gallo-Roman city of Lutetia was not absolutely confined to the island. That was the stronghold, but there were important buildings outside of it, especially to the southward. The stronghold on the island was not fortified in the early Roman time; a wall of defence was built round it only in the beginning of the fifth century after Christ. There were at least two great Roman palaces, one on the island where the Palace of Justice now is, and another on the mainland of vast dimensions, the west end of which was situated in what is now the garden of the Hôtel de Cluny. That in the island had a sort of open gallery or colonnade on the river-side;

and there is curious evidence, in some of the columns which have been recovered, that the boatmen were allowed to make use of them to haul and fasten their craft, for near the bases we find deep grooves worn by the ropes. That this Roman palace contained large rooms was proved beyond a doubt when their foundations were laid bare during the modern alterations in the Palais de Justice. The discoverers were even fortunate enough to come upon painted decorations, a specimen of which they were able to remove from the wall, and it is now preserved in the museum at the Hôtel de Cluny. Little more than this is now known about the Roman palace on the island. As its site was used long afterwards for royal dwellings, the Roman building itself may have been preserved for a long time, and have undergone a long series of alterations before it was finally replaced by a Gothic one. There have been great changes in the island since Roman times. There were no buildings in Lutetia to the westward of the palace, as its gardens went to what was then the western extremity of the island. They are now covered by the Prefecture de Police. In the times of Lutetia, and for centuries afterwards, the island came to an end in what is now the widest part of the Place Dauphine, and there were two smaller islands side by side beyond that, which have since been joined to the large one. The narrow end of the Place Dauphine is on one of these islands, and the houses on the left (as you look down the river) are partly built upon the other. There was also a long, narrow strip of an island on the left side of the larger one, and the narrow channel which isolated this strip of land has since been filled up, so that the great island has annexed three islets in all. It has also been considerably enlarged by quays built out into the river, especially at the east end, where much ground has been gained towards the Pont de l'Archevêché and the Pont St. Louis. The south side of Notre Dame is built upon the Roman wall, which it follows irregularly. The Forum is supposed to have occupied ground under the present barracks of the Republican Guard. Lutetia had one bridge over the narrow arm of the Seine, and another over the wider, but that was all. At present the island is connected with the mainland by ten bridges, if you count the Pont Neuf as two, because it crosses the two arms of the river.

Nobody knows who built the great palace to the south which bears the name of the Emperor Julian, and has long been called *Les Thermes*. Some important remains of this are still visible and are likely to be preserved, being classed as historical monuments. The great hall, which every visitor will remember, and which used to be the frigidarium of the baths, is one of the most impressive Roman remains still to be seen out of Italy. It is extremely plain, except the sculptured prows of vessels from which the vault springs; but in Roman times its broad and simple surfaces of wall and vault would no doubt be covered with stucco and decorated with some kind of mural painting, and there must have been a marble floor. It is curious that we who erect much larger buildings (though the size of this is

THE FRIGIDARIUM OF THE ROMAN BATHS, CALLED LES THERMES.

Lutetia.

considerable) should be, as we are, so deeply impressed by the power and magificence of the ancient Romans when we enter it; but this may be attributed to its antiquity. An Englishman first coming to it from England feels as an American may feel in a mediaeval cathedral; all the buildings he has ever entered are things of yesterday in comparison with this. There is something, too, which commands our admiration in the resistance to ill usage as well as to mere time. The place has been stripped bare. It has even been made to carry a garden on the top of it, and has been used as a storehouse for merchandise; yet still it stands, firm and strong, and sure to outlast all the delicate Gothic chapels in France unless they were constantly repaired. The other remains of the baths, without being so well preserved as the great frigidarium, are still sufficiently so to permit detailed recognition. The hot and cold baths, the swimming-bath, the aqueduct, the place for the heating apparatus, are all visible. It is believed that their preservation was due for a long time to the persistence of Roman customs among the Christianized Gauls, including of course the luxurious and cleanly custom of bathing according to the rules of art.

Besides what remains of the baths, three rooms belonging to the ancient palace are still in existence, and are used as part of the Cluny Museum. The lost vaults of the two larger ones have been replaced by modern roofs, but the small room is still entire. The foundations of a part of the Roman palace still exist under the Hôtel de Cluny.

"An inscription," says M. Lenoir, "placed in the courtyard of the old convent of the Mathurins commemorated the discovery of Roman remains in continuation of those under the Hôtel de Cluny, and marked their extent almost to the monastery. On the Rue des Mathurins the discoveries have been extensive, and include — 1, a great room twelve mètres square, which has lost its vault (this is annexed to the Hôtel de Cluny); 2, the understructure of two great rooms, fifteen mètres by eight, running parallel from north to south; 3, a larger room than any of these, measuring twenty-four metres by twelve. Its northern extremity (between two buildings which still exist) is ended by a curved wall like that of a Roman basilica. Possibly it may be what remains of the *consistorium* mentioned by Ammianus Marcellinus."

It is beyond the province of this little work to follow out archaeological discoveries in minute detail, but enough has been said to show that the southern palace was a building of great importance. It is believed to have been destroyed by the Normans in the ninth century.

Like other great cities of Roman Gaul, Lutetia had her amphitheatre. The ruins of it remained down to the twelfth century, or were mentioned at that time. Since then there survived a vague tradition about its locality, but all doubts were set at rest when in 1869 an important new street was cut on the south side of Paris, the street now called the *Rue Monge*. The workmen laid bare half the foundations of the amphitheatre, and the other half still remains under the modern houses. Much to the grief of the antiquaries, that half of the amphitheatre which was exposed to view had to be destroyed to make way for the modern improvements.[1] From the antiquarian point of view such regrets are quite intelligible, but from that of art the loss is imperceptible, as the remains were too low to have any architectural effect. Had the amphitheatre been as well preserved as that of Nîmes, it would have been an object of great interest, and a most valuable contrast to the monotony of modern streets. There is some reason to believe that the amphitheatre was so arranged that it might serve also as a theatre, and its western seats would be supported by the rising ground of the hill Lucotitius, that on which the Pantheon is now situated, as the seats of the theatre at Augustodunum were supported by the hill now occupied by the little seminary. In the imaginary view of Lutetia by the architect Hoffbauer the upper portion of the amphitheatre is visible on the left bank of the Seine, not very far above the upper extremity of the great island. Like the amphitheatre of Augustodunum, it would be almost out in the country.

Very little is known about the temples. Unlike Athens, Rome, Vienne, Nîmes, and a few other cities of great antiquity, Lutetia has not left a single temple standing, nor have we authentic data from which to construct a drawing of any temple that once existed. We know that there were two temples on Montmartre, one dedicated to Mars, the other to Mercury. A great piece of wall belonging to the latter existed so late as

[1] The last news is that the other half of the amphitheatre is in danger of sharing the same fate.

1618, when it was blown down by a tempestuous wind, and "the idol reduced to powder." All that we know about its shape is that it was "a great ruinous piece of wall." It is represented as such in the distance of a picture painted in the fifteenth century for the Abbot of St. Germain des Prés, and now in the Musée des Monuments Français.

Still, if we have not accurate data concerning the temples of Lutetia, we have clear evidence in the quantity of rich architectural fragments which the disturbed soil of Paris has yielded up that the place contained buildings of considerable magnificence, as did the other great Gallo-Roman cities. Lutetia seems so remote from us that we hardly realize its existence. It is more like a poetical dream for us than that which was once a reality. This is due in part to the total abandonment of the name, and in part to the nearly total effacement of all material vestiges. The case may be understood in a moment by supposing a similar effacement at Rome. Suppose that the Coliseum had simply disappeared long ago, that every vestige of temple, palace, forum, triumphal arch, monumental column, and ancient wall, had also vanished; finally, imagine a new city where Rome had been, but so big as to cover its environs, and that this new city, instead of being called Roma by the Italians, was called, let us say, Avezzano or Pescino, and had itself a more famous history than any other modern town,—what would be the consequence? Simply, that the sites of old Rome, instead of being familiar to all tourists, would be a matter of dubious speculation for melancholy-minded archaeologists, who would continually deplore its disappearance, and that the new city would go on with its business just as if ROMA had never existed. Such has been the fate of Lutetia, once a fair city, with busy commerce by land and water, with palaces, villas, aqueducts, and baths, now a dream as remote from us as Troy, the only difference being that, as we go down the Seine and pass the most historical of her islands, we know that once Lutetia was there.

In M. Hoffbauer's drawing of Lutetia the city is prudently placed at a distance, while the aqueduct of Arcueil (of which the details are known) occupies most of the foreground. We have not ventured to attempt a restoration of Lutetia seen near, so we give, instead, the view of the island as it is to-day, seen from the windows of the Louvre, certainly one of the finest urban views in the world. It has already been explained in this chapter that the great island has been lengthened westwards, that is, towards the foreground of the etching, by the annexation of two small islands, which in ancient times were separated from it by narrow channels. The elongated island now finishes prettily with a clump of trees, behind which the reader may recognize the equestrian statue of Henry IV. on its pedestal. Immediately in front of the statue are two massive blocks of houses, built in Henry's time, and remarkable for their heaped-up, picturesque, and richly varied roofs, which have often been sketched by Parisian artists. These houses are at the narrow end of the Place Dauphine, and the space between them used to be its only entrance and exit. The bridge in the foreground (I need hardly observe) is the Pont Neuf, and after it, as we look up the river on the broad arm, we see in succession the Ponts au Change, Notre Dame, d'Arcole, and Louis-Philippe. Near the Pont au Change are the mediaeval towers of the Palace of Justice, and that is the place where the Gallo-Roman boatmen, the *Nautœ Parisiaci*, used to fasten their barges to the colonnade of the Roman palace. The principal existing beauties of the island, as seen from the western extremity, are the towers of Notre Dame and the elegant spire of the Sainte Chapelle. The work of modern times has not been by any means entirely hostile to its beauty; for if the island has lost something in the vanished Roman palace and other buildings, it has gained immensely in recent times by its beautiful bridges and quays. The view was blocked in the Middle Ages by the houses upon the bridges. We shall see later how superior the modern bridge is to the mediaeval one, and what an incalculable gain the new kind of bridge has been to city views. Let us, however, always exempt from praise the modern railway pontifex, who thinks nothing of spoiling a great capital with his cast-iron abominations. To understand the injury that may be done by them, the reader has only to imagine one of them in the place of the Pont Neuf or the Pont au Change.

It has been said that Lutetia was walled late (about

Lutetia. 27

the close of the fourth century), and this first defence lasted a considerable time. It is believed that it was still in existence (probably after considerable repairs) in the time of Charles the Bald, and that ruler strengthened it by wooden towers, — one at the western end of the city, called *la tour du Palais*, and the two others at the ends of the bridges, where they abutted on the mainland. To save the reader the trouble of a reference, we may add that Charles the Bald reigned from 840 to 877. After this we know very little about the fortifications till the reign of Louis VI. (1108–1137). That monarch built two gateways in stone to defend the access to the two bridges from the mainland to the island, probably on or near the sites where the wooden towers of Charles the Bald had been, and he called these *Le Grand Châtelet* and *Le Petit Châtelet*, names which the reader is requested to remember, as they are of much importance in the topography of Paris. Etymologically, *châtelet* is exactly the same word as *chalet*, and merely means a small castle; but by one of those distinctions which custom creates between words of like origin, châtelet means a small strong castle, a work of fortification, while chalet only means the diminutive of a fine house. The present reminders of the Grand Châtelet in Paris are the Place and the Théatre du Châtelet. So little warlike is its present aspect, that the pretty square has its own theatre on its western side, and the Théatre Lyrique on its eastern, and between the two is a fountain with a column opposite an elegant undefended bridge. The extremely peaceful aspect of things inside Paris tempts us to forget that the town is still a fortress, the only difference being that its defensive castles are now called forts, and are at a distance in the country.

The Grand Châtelet had no doubt a fine imposing aspect when first built, with its lofty conical-shaped towers and gloomy portal. Our engraving shows it as it still existed, injured both by diminution and addition, in the middle of the seventeenth century. The reader will easily see how little the original military architecture had been respected. In the structure between the towers, which ends as a belfry, were the arms of Louis XII. As the work of Louis VI. had been so little respected, the complete destruction of it in 1802 need not awaken in us any very profound regret.[1]

The Gallo-Roman wall is counted by French antiquaries as the first wall, — *la première enceinte.* It is rather important to remember the order of the successive rings of wall that enclosed Paris as it grew larger, for they constantly recur in the topography of the place. The second wall was that of Louis VI., the builder of the two Châtelets; but the learned do not seem to know very much about this wall positively. They know, however, that it included much of the town which had spread out of the island, and therefore that it was the first clear definition of mediaeval Paris as distinguished from the antique Lutetia.

[1] The Petit Châtelet was a simpler building than the other, — a sort of donjon tower, with bartizans. We may have to recur to it on a future occasion. It was used as a prison. The Grand Châtelet was at one time the Provost's residence, and it became a court of justice.

THE GRAND CHÂTELET.

The third wall was that of Philippe-Auguste, and of this we know a great deal, — almost as much as if we had actually seen it. That great and energetic sovereign was as enterprising in building as in politics, and the same instinct which made him enlarge and strengthen his kingdom led him at the same time to enlarge and strengthen his capital. He boldly included in his new wall not only existing streets that lay outside that of Louis VI., but also great spaces of garden-ground, of vineyards, and even fields, which he foresaw would be covered with houses in course of time. His wall was a thoroughly good and substantial piece of work, and handsome, too, in the simple beauty of mediaeval military architecture, which, though not so rich and elegant as the ecclesiastical or domestic architecture of the same period, was still incomparably superior in appearance to the ugly military works of our own time. The *enceinte de Philippe-Auguste* consisted of two walls faced with ashlar, one facing towards the country, the other towards Paris, and the space between them was filled with cemented rubble, of which were also the foundations. The wall was three mètres thick and nine high, including the parapet, which was embattled; and at intervals of about seventy mètres there were round towers half buried in the wall, yet projecting from it about two yards: these were at first covered with conical roofs, but they were afterwards embattled like the parapet. I am not sure about their height, but suppose it to have been thirteen or fourteen mètres to the eaves of the conical roof. At longer intervals were large gates, flanked by towers of more important size, and these were fifteen or sixteen mètres high.

On the south side of the river the wall of Philippe-Auguste, which was interrupted by the Seine (there being no fortified bridge in continuation of it), started from the Tour de Nesle, which remained long after the wall itself had disappeared, — long enough indeed to be drawn and etched by Callot. This famous Tour de Nesle was originally called after Philippe Hamelin, a provost of Paris, and the name was afterwards changed when it belonged to Amaury de Nesle. It is one of the most important points in Parisian topography, and is easily remembered in connection with Callot's etchings and other prints. It is remembered also in connection with the terrible legend of a vicious queen (Jeanne de Bourgogne, wife of Philippe le Long), who is said to have enticed handsome youths into the tower and then had them cast into the Seine before daybreak that they might tell no tales.[1] We do not see the tower in Callot's representations of it quite as it was originally built. At first it is believed to have had a conical roof, and the turret staircase was added by Charles V.

The exact situation of the Tour de Nesle was where the eastern or right wing of the Institute stands at the present day.

[1] This is one of the best-known popular legends in France, being at the same time romantic and horrible, and therefore exactly suited to the popular taste; but I have very little faith in the truth of it, because, as a general rule, the water was too shallow at the foot of the tower for such deeds to pass unperceived. If done at all, it could only be when the Seine was in flood.

THE TOUR DE NESLE. FROM THE ETCHING BY CALLOT.

The reader is now requested to transport himself in imagination across the river till he is in the courtyard of what is now the old Louvre, the great square courtyard of the palace. Let him stand, in imagination, precisely in the very centre of that square and look southward, or towards the Seine. If the past could rise like a ghost he would see a phantom wall crossing the courtyard from north to south just at his left hand, and there would be one of its round towers just within the court on the north side of it near to the present entrance from the Rue de Rivoli. That would be the wall of Philippe-Auguste exactly in its old situation. Just at the same spectator's right hand would be one of the corner towers of the Castle of the Louvre that Philippe-Auguste erected. It was a square castle with a courtyard in the middle of it, and in the court there stood a great keep or donjon. The castle cannot have been of very vast dimensions, as it occupied not quite one quarter of the present square, including the site of the present building, and not simply the open space. It was, however, a strong place according to the military requirements, of the time, and is not to be confounded, in its origin, with the palatial associations that have since gathered round the word "Louvre." It began by being purely and simply a fortress, and a part of the defensive arrangements made by Philippe-Auguste. Afterwards Charles V. heightened and embellished it, opened windows in its grim walls, and turned it into an agreeable royal residence.

Now, if the reader will suppose that he is walking

from the centre of the Louvre Square straight towards the river, he will just pass on his left hand, before coming to the present quay, the site of an old tower belonging to the fortifications of Philippe-Auguste, and which used to be called *La Tour qui fait le Coin*. That tower may be seen still in old drawings, and it stood exactly opposite to the Tour de Nesle. A chain was carried across the Seine there to bar the passage.

These archaeological details may not appear at first sight to belong very closely to our subject, which is the aspect of Paris, for these towers and the entire wall of Philippe-Auguste have long since been swept away; but the Paris of old engravings is not to be understood at all without some knowledge of the past, and nothing adds so much to the interest of the present ground as the knowledge of what stood there formerly. The old court of the Louvre is a wonderful and magnificent enclosure, but the interest of it is much augmented when we know that a strong mediaeval castle once stood there, and that the city wall once traversed the same space. The Institute is a building of some architectural merit, with many noble intellectual associations; but any visitor to Paris who is cultivated enough to care about such associations as the present building possesses will probably have enough of the historic sense to care about the Tour de Nesle, and interest enough in art to know that Callot drew it. The past is interesting also for its wonderful influence in determining the sites of present buildings, often in a way which nobody would ever imagine. The visitor to Paris who knows

THE LOUVRE OF PHILIPPE-AUGUSTE.

absolutely nothing about its history is likely to imagine, when he sees the Louvre, that the site on which he finds a picture-gallery was selected for the convenient exhibition of art-treasures; whereas the truth is that it was first chosen for military reasons, when a fortress was built just outside the walls of Paris, yet near the river, and that the fortress became a royal residence, which in its turn became a national art-gallery by a series of transformations that we have still to follow. It is well to remember what has been; but there is little reason to regret the disappearance of such relics as the Tour de Nesle and the Tour qui fait le Coin. We have only to see them in drawings of their old age to perceive how incongruous and out-of-place they had become. The present Louvre is magnificent enough to deserve that the past should be sacrificed to it. Let the past be sacrificed then, but not forgotten.

III.

A VOYAGE ROUND THE ISLAND.

IT is wonderful how much the interest of a piece of land is augmented by the simple fact of its being surrounded with water. The reason probably is, that the isolation of the land gives it unity and limits, which are the first conditions necessary to every work in the fine arts. Our own faculties are so limited that the infinite always disconcerts them; but give us something so defined that we can see its boundaries, and we have the comfortable sensation that perhaps we may understand what lies within them. This feeling about islands is naturally in inverse ratio to their size. Australia, though strictly just as much an island as the Isle of Man, is never spoken of as an island at all, and we do not think of it as one. The two Americas are one island, or two peninsulas; but we call them a continent. Even Great Britain is too large for us to feel its insularity unless we think about it. The perfection of an island is to be just big enough for some variety of hill and dale, and yet so little that the whole circumference of it can be seen from some elevated point.

There are many such spots of earth in the world, of great natural beauty, in lakes, rivers, and seas; but if we except the half-artificial islets on which Venice is built, there is not an island anywhere to be compared for human interest to that which is crowned with the towers of Notre Dame and the spire of the Sainte Chapelle. What may have been its natural beauty in prehistoric times we can only guess. It has no hill, no rock, like that at Decize in the Loire. Probably it was never anything better than a flat piece of land adorned with groups of trees and reflecting itself, like hundreds of other river islands, in the stream that washed and undermined its banks. Man took possession of it, and gave it an interest surpassing that of rocks and foliage. In itself it is now nothing but a flat area, defended from the destructive action of the water by well-built quays; but every inch of it has its history, and besides this the island has an architectural interest of a peculiar kind, for the work that has been done in it in past ages, and for the remarkable changes that have been made in it both in modern and in older times.

I must now ask the reader to accompany me in a boat voyage round this famous little island, — a slow voyage, with many pauses, as different as possible from a trip in one of the swift little steamers that dart so frequently under the bridges. They are not for us. Neither do we require a swift and elegant rowing boat, such as they build now down at Asnières. Anything that will float and be steady is good enough for us; but we require an experienced *marinier de la Seine* (a worthy successor of the ancient *Nautae Parisiaci*) to look to our safety in the currents, for we shall be far too much occupied with other matters to concern ourselves about the details of navigation.

We will go down the broad arm of the Seine first, if you please, and then ascend the narrow one; and we will start from the Pont Sully, which goes from the Quai de la Tournelle across the eastern corner of the Island of St. Louis, straight in the direction of the Bastille, which the pedestrian soon reaches by the Boulevard Henri IV. Before leaving the Pont Sully, we may observe that this spot where the Boulevard St. Germain joins the Quai de la Tournelle is of considerable importance in the historical topography of Paris, because the Porte St. Bernard was just precisely there; and not only was that gate in the original wall of Philippe-Auguste, but it was preserved, after undergoing a transformation, till the comparatively recent times of Louis XVI. In the days of Louis XIV. the old Gothic gate was turned into a classical arch of triumph in honor of the great king; but a piece of the old wall and two towers were left intact on the side towards the Seine, and that which stood close to the water was the Tournelle itself, from which the present quay takes its name. For various reasons it is one of the most important points in Paris. In the Middle Ages this tower was connected by a chain with one that stood opposite to it on the island of St. Louis, while on the land side the wall which started from the Tournelle in a southerly direction, and turned westward just above where the Pantheon now stands, was the boundary of the great mediaeval university of Paris. What is now called the

Island of St. Louis was in the fifteenth century two islands; the one to the east being called *l'Ile aux Vaches*, and that to the west *l'Ile Notre Dame*. Farther east, and separated from the Ile aux Vaches by a narrow channel, and by one still narrower from the north shore of the Seine, was another island, called *l'Ile des Javiaux*. This was called *l'Ile Louvier* in the eighteenth century, and was used as a storage ground for firewood; but the channel has now been filled up and the island annexed to the mainland. The Boulevard Henri IV. and three smaller streets cross what was once flowing water. As to the present condition of the Ile St. Louis, it need not detain us. The ground is covered with the usual tall, well-built, modern Parisian houses, and connected with other parts of Paris by seven bridges, if you count the Pont Sully as two, which it really is. The island is said to be an agreeable place of residence for its almost Venetian quiet, and for the fine views from many of its windows. Altogether it has now a very highly civilized appearance; yet one cannot help regretting the fifteenth century, when there was only a bit of fortress wall upon it, with towers, and a few trees, and when seventeen towers could be counted along the north bank of the Seine, and turning up to the great fortress,— the Bastille,— while within the space so enclosed arose many a turret and spire whereof there are none remaining.

The Isle of St. Louis — which in the Middle Ages had been so little dealt with by human art that the animals upon it could get to the water all round, except where the banks were undermined by the current — is now so surrounded with quays that the horses in the stables could never approach the water at all unless access were made for them artificially. This is one of those numerous cases in which civilization first takes away a natural convenience and then restores it in its own fashion. Frenchmen are very fond of bathing their horses in the fine weather; you may see them doing it in all the rivers of France, as artists are well aware. Nothing that men and animals are ever engaged in together offers prettier and more unexpected effects of grouping and active movement, while the rippling of the water itself against the animals' bodies affords ample variety of reflection. The view from the river here has been much diminished in picturesque interest by the gradual and now almost complete victory of modern neatness in the works of the house-architect and the engineer, the only very obvious gain being the distant dome of the Pantheon. Notwithstanding the loss of all the military mediaeval towers, such as the Tournelle on the left bank, the Tour Loriaux on the Ile St. Louis, and many others, we have one consolation which makes us easily forget them all. Notre Dame is still erect on the greater island, the glory of the river as you come down through the Pont de la Tournelle, so that you can hardly take your eye off it as the motion of your boat changes for you the intricate perspective of tower and spire and flying buttress. There is many a fine river-scene in France in which natural beauty is mingled with some remnant of noble architecture. Here the natural beauty is limited to sky and water, and to the trees in the space at the upper end of the island, now called the *Jardin de l'Archevêché;* but it is a scene which nothing spoils, and which has a wonderful charm and grandeur at certain times, especially in the splendor of sunset. Notre Dame looks imposing from every side; but there is no view of the building quite so impressive as that which includes the apse, with its long, light, flying buttresses in their varied degrees of foreshortening.

This illustration shows the cathedral as it appears from the garden itself; but, like all large edifices, it is much more imposing from some distance, and looks best in the well-known view from the left bank of the Seine that has been so often drawn, painted, and engraved, and that was the subject of one of Méryon's most famous etchings.

The contrast between the two islands of St. Louis and La Cité is in nothing more remarkable than in the antiquity of the human life upon them. Here the reader must be requested to give his special attention for one moment to one of those points which are the perpetual confusion of the careless and unobservant. When the careless reader meets with the Ile Notre Dame in the history of Paris, he inevitably imagines that it is the island on which Notre Dame is built; whereas it was the mediaeval name for the more southerly of the two islands, now united into one under the name of St. Louis; and what is most curious and remarkable is, that although the island of the city on which Notre Dame is situated was peopled in the time of the

Romans, and covered with a most dense population in the Middle Ages, the island called after Notre Dame was waste land until the seventeenth century. This accounts for the strange fact that there never was a mediaeval bridge from one island to the other, though they are so near that a bridge seems inevitable. The distance is only sixty-five mètres, and it is now spanned by a single arch. In the seventeenth century there was a wooden bridge from the southern extremity of the Ile St. Louis (the present Pont St. Louis is higher up), and this wooden structure has a strange history connected with what was called the Cloister of Notre Dame. This was not what we are accustomed to call a cloister, but a sort of ecclesiastical village composed of thirty-seven houses, each having its own garden, and the whole being defended by a wall. The cloister was situated in the Island of the City, between Notre Dame and the channel now crossed by the Pont St. Louis. It appears that the clergy who lived in it enjoyed such delightful quiet amid their gardens that they could not endure the idea of a bridge with its noisy traffic; so in order to spare the cloister the bridge was made of a very peculiar form. First it crossed the channel at such an angle as to make it much longer than necessary; and then, when it had got near what is now the Quai Napoléon, it ran parallel with the shore of the island for some distance before landing. This wooden bridge is known in history as the Pont Rouge, because it was painted with red lead.

Next we come to the Pont d'Arcole, which has

GARDEN EAST OF NOTRE DAME.

hardly any history.[1] It is in one arch, and a light and clever piece of modern engineering. It connects the Rue d'Arcole, which leads to the west front of Notre Dame, with that part of the north shore where the Hôtel de Ville is situated; consequently in modern Paris there are few points of greater architectural interest. Still, so far as the variety and abundance of picturesque material is concerned, this part of Paris has suffered even more than many others by modern improvements. It was once extremely populous. In the Middle Ages it was a labyrinth of narrow streets, with tall gabled houses, all along the bank of the river. Even in the last century there still subsisted a number of small churches and tortuous streets, many of which bore the old names, and remnants of them may still be remembered. The improvements, begun under the reign of Napoleon III., and carried out under

[1] I quote the following passage in a letter from my old friend William Wyld, the distinguished painter, as it adds to the interest of this bridge: "Touching the Pont d'Arcole of which you say, I think, that there is 'hardly any history,' I will tell you a little anecdote. In 1830 I was an eye-witness to much hard fighting across that bridge (which was then but a small suspension *passerelle* for foot-passengers only), and saw there many a tall fellow laid low (I was on the quay of the Isle St. Louis). I think the bridge was then called the *Pont de l'Hôtel de Ville*,—but during the hottest of the fight a youth dashed sword in hand on to the bridge crying out, '*Je m'appelle d'Arcole!*' ... Whether he escaped or not I don't know, but his name was given to the bridge without any allusion to that in Italy on which Napoleon I. carried the flag of the Republic amid a shower of bullets." According to Joanne's Guide, the youth was killed upon the bridge. Certainly nothing was subsequently heard of him; and so by a single exclamation joined to an act of courage he won lasting fame, slightly obscured only by the confusion between his name and that of a bridge in Italy. Joanne gives his name as " Arcole," without the particle.

the Republic, have cleared away all these, and substituted for them broad streets, enormous public buildings, and an extensive open space. The result has been a simplification not unfavorable to the effect of magnificence, but very destructive of the picturesque, because the picturesque requires the variety of many unexpected details. The pyramids of Egypt are grand, but not picturesque; the streets in old Cairo were picturesque in the extreme. The new Hôtel Dieu, which has one front to the Seine and another on the open space in front of Notre Dame, is so vast that the site of it covers nearly three times the extent of ground occupied by the cathedral. On the site of this single building there used to be three churches and part of a fourth, and no less than eleven streets![1]

Another modern taste besides that for extensive public buildings has been extremely destructive of houses. Throughout the Middle Ages, and even down to comparatively recent times, it was considered a wise economy of space to cover bridges with houses; and that to such a degree that instead of a

[1] As some of these have a certain degree of historical interest, I give the names of them in a note. The churches were those of St. Landry, St. Denis de la Chartre, La Madeleine, and part of St. Pierre aux Bœufs. The streets were Rue Basse des Ursins, Rue Haute des Ursins, Rue du Haut Moulin, Rue des Marmousets (celebrated for the pastry-cook who in the Middle Ages made pies of human flesh, and etched by Lalanne before its demolition), Rue du Chevet St. Landry, Rue St. Pierre aux Bœufs, Rue Cocatrix, Rue de Perpignan, Rue des Trois Canettes, Rue de la Licorne. This list may serve to give the reader some faint idea of the enormous effacement of old Paris which has been necessary to make room for the gigantic modern public buildings, all this being sacrificed to a single hospital.

broad road over the bridge with open views on both sides of it, the people of those days had the advantage (as it must have seemed to them) of getting across the water through a narrow street without any views at all, but with plenty of delightful little shops. The old notion of a bridge was to have it as much as possible like the present *passages* of Paris, such as the Passage Jouffroy, the Passage des Panoramas, etc. It may be supposed that our ancestors thought a bridge without houses bleak and uncomfortable, as the passengers over it would be unpleasantly exposed to the draught of wind that generally blows up or down a river. Their arrangement gave no view, but it gave a sheltered lounge, with plenty to see in the shop-windows. A curious consequence of it, which scarcely strikes us until we reflect a little, was that not only were the passages deprived of a view on the river, but even from the windows of the houses themselves very little was to be seen, as the *next* bridge always blocked the view, so that when the bridges were near together the houses on them and on the banks of the river made a sort of square with an enclosed area of water; and the river was little better than a succession of such squares. We may be severe on our own times for some errors of taste, but surely in our treatment of rivers we have reason on our side. The Middle Ages had nothing to show like the quays and bridges of modern Paris. There was not a single spot in the Paris of Philippe-Auguste from which a view could be had up and down the river like the view from the modern bridges and quays. Old Paris had a thousand picturesque bits, but it had no distances.

The Pont Notre Dame is that which joins the present Rue de la Cité, which is on the island, to the Quai de Gèvres on the right bank of the Seine. It is one of the most interesting bridges in Paris. I have not space to give the history of the earlier bridges, and the reader might not care to follow such archaeological details; but we cannot pass in silence the wonderful catastrophe of October 25, 1499, when the bridge fell into the Seine with all the houses upon it. The year previous some carpenters had noticed the rotten condition of the piles, and gave ample warning; but this was disregarded till at length a master carpenter went to one of the authorities, the "lieutenant-criminal" Papillon, and told him that the catastrophe was imminent. By order of a court then sitting (at seven o'clock in the morning) Papillon went and gave notice to the inhabitants and closed the bridge to the public. The dwellers on the bridge tried to remove their goods (a great piece of labor as they were all shopkeepers), but could not effect this before the entire structure fell into the river with a fearful noise, and amid such a cloud of dust that nothing could be seen. It is difficult to imagine anything more terrible except an earthquake. In the midst of the confusion some lives were strangely preserved. A porter with a burden of arrows on his back was thrown into the river and simply swam to the side. A man in one of the houses seeing a fissure yawn beneath him jumped out of window and also saved him-

PONT NOTRE DAME, 18TH CENTURY.

self by swimming. But the most remarkable case was that of a little child, tied up closely in its swaddling clothes and lying in its cradle. The cradle was flung into the water, where it was afterwards found floating like a boat, with the child alive and well inside it: so, at least, says a contemporary chronicler.

The custom of having houses on bridges was too deeply rooted for the new one to be without them, so it was covered with tall structures, with their gable-ends to the stream, — more than thirty gables, like the teeth of a saw, according to a careful old engraving. The bridge so erected at the beginning of the sixteenth century remained essentially the same till the eighteenth, except that the fronts of the houses were modernized according to the taste of the day. A curious point to be noted is that these houses were the first in Paris to be numbered, and with odd numbers on one side and even numbers on the other. It was a place for fashionable shops, kept by jewellers, goldsmiths, picture-dealers, — a sort of Palais Royal or Boulevard des Italiens of that time.

This street on the bridge existed till towards the close of the eighteenth century, when Louis XVI. decreed the demolition of the bridge-houses throughout the capital. This innovation was very nearly contemporary with the political revolution, and was first carried into effect on the Pont Notre Dame. Wonderful to relate, in spite of so much modern improvement the old bridge still exists; but it has been re-cased in stone and altered in some respects externally, so that it has now quite a modern air.

The very last relic of old-fashioned picturesqueness about the bridges of Paris was the pump just below the Pont Notre Dame, built originally in 1678. I remember it well; and not only do I remember the thing itself as a material object, but also a certain feeling that it awakened, — a feeling of respect for a sort of majesty that the poor old structure undoubtedly possessed, and of regret that the march of improvement would so soon remove it. Méryon made a delightful etching of it, one of the most remarkable of all his plates for clearness and elegance of style, and he also wrote some verses in pity for its fate. His etching showed the pump in afternoon light; the accompanying woodcut shows the aspect it had on sunny mornings. The truth is that, though a poor, cheap structure, it had several fine architectural qualities. Its masses were well composed, well supported, and admirably crowned by the tower.

The short space between the Pont Notre Dame and the Pont au Change is one of the most interesting in Paris. The flower-market, as pretty a sight as the modern city could show anywhere, used to extend in an open space between the Quai Desaix and what was the Rue de la Pelleterie. It had a fine background towards the west in the buildings of the Palace of Justice, with the picturesque corner tower, and it inspired artists with the desire to make pictures of it.[1] I wonder what artist would care to paint the same scene to-day. Instead

---

[1] A drawing of it by Turner was engraved in the "Rivers of France," but it is one of the weakest in the volume. Turner especially missed the character of the clock-tower, which is and always was very definite and peculiar

THE PUMP NEAR THE PONT NOTRE DAME, 1861.

of the pleasant open space which so charmingly disengaged the buildings of the Palace, we have now a great, heavy, ornate, and vulgar modern edifice with a dome, — the Tribunal de Commerce; just one of those erections which the Philistines always consider "very handsome," and look upon with deep respect because of their evident costliness. The best time of this bit of Parisian scenery, from the artist's point of view, must have been when Girtin drew it in 1800. Then the towers of the Palace were not yet united by heavy masses of modern building which reduce their importance, and the old corner tower had not been replaced by the new one. If we go farther back the view is spoiled again by another cause. During the latter half of the seventeenth and nearly the whole of the eighteenth centuries a massive line of stone houses, five stories high, stood on the Pont au Change, and of course effectually blocked the view. Earlier still a row of gabled post-and-plaster houses stood on a wooden bridge, but these were all burnt down in 1621. There was at that time another bridge, a very little lower down the river, called the *Pont Marchand*. A servant-girl there let a candle fall in a place where firewood was kept, probably among shavings, for the house was soon on fire, and with it the others and the bridge itself. The flames soon reached the Pont aux Changeurs, which was totally destroyed. Not only were the bridge-houses burnt, but some on the land caught fire also. An eye-witness has left an account of this fire, which must have been a most remarkable spectacle. When the houses were already in flames the inhabitants remained as long as possible, throwing their goods out of the windows. There had been warning for the destruction of the Pont Notre Dame: for this there was no warning.

The houses between the two bridges on the side of the island had gables and balconies towards the river; the lower stories of them, near the water (there being no quay in old times), were occupied by tanners who congregated in this one quarter, according to the mediaeval custom. If these buildings could have been preserved to our own day, they would have been favorite subjects for Parisian artists (for whom there is little left); but as the natural progress of a modern city is towards good quays, all the humble old river-side industries have to go elsewhere.

I mentioned the simplification which had resulted from modern improvements on the island north of the Pont Notre Dame, where eleven streets and three churches had made way for a single building. The same process has been carried out in the section of the island which is included between the lines drawn across it from the Pont Notre Dame and the Pont au Change. In this area there were formerly nine streets and four churches,[1] but at the present day there are simply two buildings, — the Tribunal de Commerce, already mentioned with the degree of respect due to it, and a huge barrack called

[1] The streets were Rue de la Pelleterie, Rue Gervais Laurent, Rue de la vieille Draperie, Rue St. Eloi, Rue de la Calandre, Rue aux Fèves, Rue des Carcuissons, Rue du Marché Neuf, Rue St. Croix; the churches were St. Barthélemi, St. Croix, St. Eloi, and St. Germain. All these, as well as the Marché Neuf, have entirely disappeared.

the *Caserne de la Garde Républicaine*, of which we can only say that it is very extensive, very well built, and as tiresome as it is extensive. The real improvement which has followed from recent changes is, that the few modern streets — the Rue de la Cité, the Boulevard du Palais, and the Avenue de Constantine — are so much more spacious than the many little streets of former times, and give such superior views. They are three or four times as broad as the old Rue de la vieille Draperie.

In the Middle Ages there was a bridge for footpassengers only, but with houses upon it, just below the present Pont au Change. This was erected exclusively for the convenience of the millers, who were allowed to occupy nearly the whole width of the river with their wheels, placed in the open spaces between the wooden piles of which the bridge was built. The whole structure was carried away by an inundation towards the close of the year 1596, and it was afterwards replaced by the Pont Marchand, destroyed in the great fire of 1621. There were eleven mills, and the names of the millers have been preserved. After the fire it was not thought necessary to rebuild the lower bridge, which was not of much public utility, so the Pont au Change has remained by itself ever since. The present structure is of very recent date, having been built in 1859, not quite in the same angle as the old bridge (being now more at right angles with the river), and a little higher up the stream, especially on the side of the island. There is little to be said about its architecture, except that the essentially modern ideas of depressed arches and level roadway have been carefully adhered to, while a certain elegance is given by a cornice and balustrade. Such is the course of bridge-architecture from the Middle Ages to our own time. First comes the wooden mediaeval bridge, consisting simply of tall piles rising straight from the bed of the stream, and bearing a street of crowded houses upon them; next, the substantial, round-arched stone bridge of the sixteenth and seventeenth centuries, burdened with stone houses, massive and lofty; then the same bridge without the houses; and, lastly, the modern bridge, with depressed arches of wider span, and a broad, level roadway above. The modern ideal is by far the most rational of all, being at the same time the most convenient for vehicles above and boats or rafts below, while it reduces to a minimum the obstruction of the view. The only objection to it is, that its extreme simplicity of purpose has a tendency to produce a merely utilitarian structure, unless the architect is a man of great taste and intelligence, who can give a touch of elegance to a work of plain utility.

There is a well-known etching by Méryon showing the Pont au Change and the round towers of the Palace of Justice, seen through an arch of the Pont Notre Dame, with the wooden substructure of the old pump to the spectator's left. This etching gives, as well as any existing illustration, the character of the old Pont au Change with its round arches, its plain parapet, its rising roadway, and its angular cutwaters. The plate is interesting, too, for the ingenious introduction of the round

THE PONT NEUF IN 1845.

towers, which are now all that is left of the picturesque between Méryon's position and the Pont Neuf. On the right bank of the Seine you have two pretty theatres (Châtelet and Lyrique), and the light column of the palm-fountain, with the very elegant tower of St. Jacques, more visible than ever before; but the picturesque of the river-side is gone.

In the Middle Ages there was no bridge connecting the island with the mainland farther west than the Pont Marchand. Another communication was felt to be desirable long before there was a definite project, and the project was under consideration long before it was executed. The work of the Pont Neuf was at length actually and practically commenced in the year 1578, under Henri III., by driving piles on the south side, and the southern half of the bridge — that across the narrow arm of the Seine — was completed long before the other. Henri IV. took up the work vigorously in 1598 and finished it in 1604.

Old fashions linger long, and although no houses were erected on the Pont Neuf, small wooden booths were tolerated upon it for a long time; and after they were removed, they had descendants even in the present century in the shape of curious little semicircular shops erected on the projections between the arches. These are still visible in Méryon's beautiful dry-point of the Pont Neuf. They have since been removed, and the present aspect of the bridge very closely resembles its aspect in the seventeenth century. The woodcut opposite page 50 shows the shorter portion of the bridge, — that over the narrow arm of the Seine as it appeared in 1845, the little shops being still visible as turrets not very disadvantageously.[1] Turner liked them, certainly, as they are made quite prominent in his impressive drawing of the entire bridge, while he would certainly have removed them if they had displeased him. On the contrary, he was so delighted with them that he made them three times as big as they are in reality, relatively to the width of the arches.[2]

We will now, if the reader pleases, turn up the narrow arm of the Seine till we come to the Pont St. Michel. There is a particularly fine view from this bridge, of which Lalanne made a very successful etching many years ago. The beauty of it consists chiefly in the distance, which shows the long perspective of the Louvre immediately above the Pont Neuf. All the bridges on the island except the Pont Neuf have been carried away by some disaster; but that famous one has become a proverb for a sound and lasting constitution, so that robust Frenchmen proudly compare themselves to it, and complimentary ones apply the comparison to their friends (never to their political opponents, who are always represented as unhealthy). Few bridges have been more unlucky than the Pont St. Michel. In the fifteenth century it was carried away by a pack of

[1] In this view we are looking *down* the river from the Quai des Orfèvres.

[2] This assertion is founded on strict measurement. In reality the semicircular projections on the Pont Neuf measure *less* than one third the diameter of the arch. In Turner's drawing, those to the left are represented as equivalent in their diameters to the diameters of the arches under them.

THE MORGUE IN 1840.

moving ice. It was again carried away in the middle of the sixteenth century, and destroyed again in the seventeenth. The next structure was of stone, with houses; the present one was built by that true *pontifex maximus*, Napoleon III.

Close to the Pont St. Michel, on the island shore, used to stand a famous little building which had at one time been a *boucherie*, and which for many years served as the dead-house for bodies found in the Seine. The "Doric little Morgue" will long be remembered on account of the immortality conferred upon it by novelists and also by at least one famous poet, Browning, and one great artist, Méryon.

"First came the silent gazers; next
  A screen of glass, we're thankful for;
Last, the sight's self, the sermon's text,
  The three men who did most abhor
Their life in Paris yesterday,
  So killed themselves: and now, enthroned
Each on his copper couch, they lay
  Fronting me, waiting to be owned.
I thought, and think, their sin's atoned.

"Poor men, God made, and all for that!
  The reverence struck me; o'er each head
Religiously was hung its hat,
  Each coat dripped by the owner's bed,
Sacred from touch: each had his berth,
  His boards, his proper place of rest,
Who last night tenanted on earth
  Some arch where twelve such slept abreast, —
Unless the plain asphalte seemed best."

The Petit Pont is of far greater historical interest than the Pont St. Michel. At the present day it consists of a single stone arch of depressed curve, and is precisely the sort of structure in which the modern engineer displays his skill; but the rather elegant and extremely simple Petit Pont of the present day has had many very different predecessors. At that spot the Romans had a bridge joining Lutetia to the mainland; and just here, where the bridge abuts on the south bank of the Seine, the gate, fortress, and prison called the *Petit Châtelet* stood grimly in the Middle Ages, and even down to the last years of the eighteenth century. It was a building of sinister aspect, with few openings to the light of day, and nothing in the way of ornament except four simple string-courses and about as many bartizans. A Gothic archway led through it from the bridge. I willingly spare the reader an account of the cruelties committed in this building, and will speak of the bridge only. Unluckly as were the other bridges of the *cité*, this was the most unfortunate of all. It is said that the rapidity of the current in flood-times was the cause of successive accidents, now happily at an end by the construction of a single arch beneath which the floods rise freely. M. Jourdain tells us, in "Paris à travers les Ages," that the Petit Pont fell in 1206, 1280, 1296, 1325, 1376, and 1393; but the most remarkable of all its misfortunes occurred much later, in 1718. At that date it consisted of a good stone bridge of three arches covered with tall stone houses; but it seems as if the contemporary engineers had not much confidence in

THE LITTLE CHÂTELET, TAKEN FROM THE PETIT PONT IN 1780.

their strength, for beneath the arches, and also at the ends of the piers, there were strong wooden scaffoldings, like those supporting the pump that Méryon drew. Now it so happened in that year 1718, in the month of April, that a woman had lost her son by drowning, and that her grief was greatly increased because she could not find his body; wherefore the good folks, her neighbors, told her of a sure method by which drowned bodies might be found, and she believed and obeyed them. She took a *sébille*, which is a thick, round wooden tray or dish, she stuck a taper upright in it, which she lighted, and with the taper she put a piece of blessed bread, the whole in honor of St. Nicholas; she then confided her little boat to the current and watched its course. It floated straight to a barge laden with hay, the taper set fire to the hay, the men in the barges near to it severed the rope that fastened it in order to save their boats; and now, instead of the little votive taper in its wooden dish, a great blazing haystack floated quickly down to the Petit Pont, where it was stopped by the wooden piles under the arch. These soon caught fire, and so did all the houses, but the fortress of the Petit Châtelet remained uninjured. The houses were never rebuilt.

There is now nothing whatever of visible interest between the Petit Pont and the upper extremity of the island, except the view of the south side of Notre Dame. Changes within our own recollection have entirely altered this part of Paris, much to its advantage. The old Hôtel Dieu occupied the whole space between the present Petit Pont and the then existing Pont au Double, which stood higher up the river than the bridge now bearing the same name; and not only did the great hospital occupy a long range of building, as ugly as a factory, on the island, but it also had another large building across the water, on the south bank of the Seine, and a block called the *Salle St. Cosme* on the bridge between them. All this effectually obstructed the view of Notre Dame; and, indeed, that half of the hospital which stood upon the island was on what is now the open space in front of the cathedral. Artists are not agreed as to the policy of disengaging cathedrals so much as Notre Dame is now disengaged; and certainly the cathedral at Rouen comes upon us with a sudden impressiveness in the midst of the narrow streets and from the small market-place, — an impressiveness which would be lost if it could be set in the middle of a large field; but Notre Dame was in former times so much injured by the vast size of the ill-contrived old Hôtel Dieu, that the removal of that particular obstruction is unquestionably a great gain. In old times the cathedral used to be hidden in a considerable degree by the *archevêché*, now entirely removed. The archbishop now lives in a fine Louis XIV. mansion in the Rue Grenelle St. Germain. The accompanying reproduction of an etching by Israel Sylvestre shows the Archbishop's Palace as it existed in the seventeenth century, and the reader may also see how the buildings of the Hôtel Dieu stretched across the river.

Nothing remains to be said concerning our circum-

THE ARCHBISHOP'S PALACE IN 1650. FROM AN ETCHING BY ISRAEL SYLVESTRE.

navigation of the island except that the eastern point of it, which in the Middle Ages was a shapeless piece of waste land called *Le Terrain*, and in the eighteenth century a garden called the *Jardin du Terrain*, is at present very neatly arranged in true modern Parisian style, and serves as a pretty site for a melancholy little structure, the new Morgue, to which the inhabitants of southern Paris have immediate access by the Pont de l'Archevêché, a bridge which, unlike its elder brethren, has no history.

A sketch of Anglers by Mr. Jacomb Hood gives a bit of topography in its background which illustrates our present subject. The anglers are on the Quai des Tournelles, the church is Notre Dame (showing the apse), the bridge is the Pont de l'Archevêché, and the bit of land going from the bridge to the spectator's right is what was formerly called *Le Terrain*, and is now well embanked and defended by a river-wall; while the low building whose roof seems to crown the wall near the boy's fishing-rod is the present Morgue.

The quays on both sides of the Seine appear to belong more to the ordinary life of the city than the more recently built embankment of the Thames. It generally happens that some idle youth may be seen lounging over the parapet and watching sympathetically an absorbed angler below who from some stair, or boat at anchor, or narrow ledge of masonry, pursues through successive hours his mildly exciting sport. It is one of the most curious contrasts in the French character that, although it is said to be impatient, and often shows remarkable irritability, it is nevertheless exactly adapted to the humblest and dullest sort of angling. Nothing can exceed the patience of Parisian anglers or their entire absorption in their pursuit. So completely do they forget everything else in the indulgence of their passion, that during the dreadful day of the Commune, the 24th of May, 1871, when the Communards were setting fire to the public buildings, and the soldiers from Versailles were shooting down the people in the streets, one or two faithful *pêcheurs à la ligne* still followed their tranquil pastime close to one of the bridges; I believe it was the Pont Neuf.

ANGLERS ON THE QUAYS.

IV.

NOTRE DAME AND THE SAINTE CHAPELLE.

THERE are absolutely only these two churches left standing in the island of the city, and there is nothing in the history of Paris which more clearly exhibits the modern disposition to make a *tabula rasa* of the past. The wonder is that Notre Dame and the Sainte Chapelle should themselves have been preserved down to our own time. There they stand, however, somewhat injured by restoration, yet happily not so much as they might have been, and likely to last for centuries still to come, considering their present excellent condition of material repair.

But where is the crowd of little churches that clustered round Notre Dame, as children round their great mother? In the Middle Ages she seemed to gather them about her as a hen gathers her chickens under her wings; but now they are all gone, and she would be left in the most complete solitude were it not that from the court of the Palace of Justice there still rises one solitary spire answering to hers, and still, as in the Middle Ages, the birds fly from one to the other.

But where is St. Denis du Pas, where is St. Jean le Rond, and where may St. Christopher, Ste. Geneviève,

St. Agnan, St. Landry, St. Peter, St. Denis de la Chartre, Ste. Marine, and the Magdalen, find the churches once dedicated to them? Can you discover even the sites of St. Luke, Ste. Croix, and St. Germain le Vieux? Have you ever seen St. Pierre des Arcas, St. Barthélemi, and St. Eloi? "There is my bridge still," Saint Michael may think; "but as for my church, I seek for it in vain!" Where are all these churches of the past, which once stood in consecrated ground, and were thought to be safe forever, — churches adorned by the mediaeval architect, often repaired and injured by later experimentalists at the Renaissance, yet interesting always for the bits of beautiful old work to be found in them? *Où sont les neiges d'antan?*

Before the present cathedral of Notre Dame there was a predecessor built by Childebert, of which we do not know very much. It occupied part of the site of the present edifice, standing near the Roman wall, and to the southeast of it there was another church dedicated to Saint Stephen. The site of the original Notre Dame is now partly covered by the west front of the edifice and a small portion of the nave, and partly left open in the space before the cathedral. It was of Romanesque architecture and of some splendor. Probably, if it had been preserved to the present day, we should have looked upon it with great interest as a very early specimen of church-building, but it is not likely that it would have produced on our minds, accustomed as we are to the magnificence of fully developed Gothic, the effect that it produced on its own contemporaries. As for the

Notre Dame and the Sainte Chapelle. 61

site of St. Etienne, the present sacristy stands on a part of it.

The present cathedral of Notre Dame was begun in 1161, the first stone being laid by a Pope, — Alexander III., — and in 1185 mass was said at the high altar. This would only prove that the choir was finished, or at least covered in. The southern entrance was begun in 1257, and the great western entrance from the Place du Parois was finished in 1223, up to the line of the gallery. The towers were finished in the time of Saint Louis.

An important matter in the history of Notre Dame is the fire of 1218, caused by thieves, because that brought about an architectural alteration clearly described as follows by M. Drumont in " Paris à travrse les Ages: "

"Before this fire the great flying buttresses of the nave and choir were constructed *à double volée*, which means that instead of crossing over the space between the buttresses and the vaults in arches of a single span, they were composed of two portions or arches, with an intermediate support. The fire probably injured the second span of the old flying buttresses. At that time other cathedrals had been erected, and their walls were pierced with larger windows, filled with brilliantly stained glass, — a decoration which was rapidly becoming important. Instead of repairing the harm done by the fire, the restorers of that time seized upon the opportunity for suppressing the rose-windows pierced above the galleries, and brought the upper windows lower, cutting away their support down to the archivolt of the galleries. The flying buttresses *à double volée* were demolished, and the height of the windows of the triforium was reduced by lowering its vaults."

The tall windows were filled with simple tracery, and in the opinion of Viollet-le-Duc the majesty of the first edifice was in a great measure sacrificed by these changes. So far as I am able to judge by M. Hoffbauer's drawing, which restores the apse to its primitive condition, and shows the double-arched buttresses, the most striking difference between the first apse and the present one was in the successive stages of roof which were visible in the first, while at present only the highest roof is visible, the others having been so much lowered in pitch to make way for the elongated windows that they are no more to be seen. The change, in fact, is that change which we find everywhere in the progress of Gothic architecture, — from a simple, strong-looking, and dignified style, to a lighter, more airy, more delicate, and elegant style. It is perfectly intelligible that a master of architecture like Viollet-le-Duc, who knew all about construction, should have preferred the first apse, with its short, plain windows, its visible tiers of roof, and its substantial, doubly supported buttresses; but, at the same time, it is intelligible that most people should prefer the east end of the church as it exists at present, with its light, far-leaping buttresses, its long clerestory windows, and the rich windows of the chapels and aisle, decorated externally with crockets and finials. Besides, there are many pinnacles now (people always like pinnacles, — the great popularity of Milan Cathedral is due to them), and it does not appear that there were any pinnacles about the first apse.

The great west front, where the towers are, is one of

the chief architectural glories of France. There is hardly any work of architecture in the whole world, except one or two Greek temples, which has evoked the same kind and degree of admiration as the west front of Notre Dame. It is considered to be one of those rarest products of consummate genius in which imagination of the highest kind works in perfect accordance with the most severe reason. May I confess frankly that until I had carefully studied it under the guidance of Viollet-le-Duc, the front of Notre Dame never produced upon me the same effect as the west fronts of some other French cathedrals of equal rank? I believe the reason to be that Notre Dame is not so picturesque as some others, and does not so much excite the imagination as they do. It is well ordered, and a perfectly *sane* piece of work (which Gothic architecture is not always), but it has not the imaginative intricacy of Rouen, nor the rich exuberance of Amiens and Reims, nor the fortress-like grandeur of Bourges, nor the elegant variety of Chartres. A man of very high architectural attainments would probably value the romantic element less than I do, simply because much that seems rich and imaginative to an amateur in architecture is understood too quickly in all its details by a master for it to produce the same poetic feeling in his mind; and I observe that architects esteem especially the judicious *ordonnance* of parts, which is a great virtue no doubt, but a very sober virtue, imposing a very strict discipline on the imagination. The truth is, that the virtues of the west front of Notre Dame are rather classic than romantic. Everything in it seems the result of perfect knowledge and consummate calculation. There are none of those mistakes which generally occur in works of wilder genius. Story after story the massive front rears itself to the towers; every division of it is acceptable either as a resting-place for the eye or as an attraction. First, you have the three great doorways, with the world of sculpture usual in the French Gothic portals, but the row of statues does not come out and round the buttresses as at Amiens and Reims. The buttresses are left plain except that there is a niche in each of them twenty feet from the ground, and one statue in each niche with its feet higher than the heads of the great statues. Above the arches the wall is perfectly plain instead of being enriched with crocketed gables, as at Amiens and Reims; and above this plain space comes the great gallery of the kings, with its twenty-eight statues in their niches. Over this gallery runs a sort of platform or balcony called the *Galerie de la Vierge;* and then we come to the great space of wall, very plain in itself, which is occupied by the great windows, the rose in the middle and the ogival windows, of two lights and a rose above, in each of the towers. Perhaps the most especially characteristic thing in this front is the light colonnade above the windows, which makes a sort of open screen in the space between the towers, and by this means prevents too much abruptness in the separation of the towers from the main mass of the building. This colonnade is not only extremely elegant in itself, but it is placed with so much judgment as to give a lightness

TYMPANUM OF THE PORTE STE. ANNE.

Notre Dame and the Sainte Chapelle. 65

to the whole front, which could hardly have been obtained by any other means. The upper part of the towers is remarkable for the great length of the openings (about fifty feet), and both towers seem to terminate very plainly and abruptly, having no pinnacles and nothing to relieve the level line except the little turrets at the top of the staircases. This, however, is explained by the fact that it was intended to have spires, and that the towers we see were entirely designed with a view to them. That project was never carried into execution, and even the enterprise of the nineteenth century shrank from it when Notre Dame was restored. Is the absence of the spires to be regretted? We have some means of judging this question by a comparison of the west front as it is with the drawing of it with spires which was engraved and published in the "Entretiens sur l'Architecture," by Viollet-le-Duc. So far as the towers only are concerned, the effect of the spires is excellent. They at once reduce the long louvre-windows to due proportions, and remove the otherwise unaccountable plainness of the summits of the square towers. But on the rest of the front the effect of the spires is not so happy. The arcade is tall enough not to be stunted by them, but the gallery of the kings and the great doorways are made to appear much more insignificant than they are at present. At the same time two considerations ought not to be forgotten. It is quite possible that the spires intended by the mediæval architect may have been lighter in appearance than those designed by Viollet-le-Duc, and it is also to be remembered that an architect's elevation always produces quite a different effect upon the mind from the sight of the reality in stone. Had the spires been completed, no one approaching close enough to see the statues in the portals and in the gallery of the kings would have seen the spires at the same time; he would only have been conscious of their existence.

I have said that the virtues of the west front of Notre Dame are rather classic than romantic. It is a generally received idea that exact symmetry was one of the classical characteristics; but a closer examination of classical works reveals unsuspected varieties in measurements which are supposed to have had for their object the avoidance of mechanical dulness. The variety in Gothic architecture is so frequently apparent that the popular mind associates the idea of variety with Gothic work as it associates symmetry with Greek. There are, however, in Gothic buildings certain parts which appear to be symmetrical, and which frequently are not so. That this variety was intentional is quite certain. An architect is not like a landscape-painter who draws by the eye, and may accidentally make one object smaller than another when he intended them to be alike. An architect measures everything, so that, so far as dimensions are concerned, there can never be an undetected error in his completed work. The two towers of Notre Dame, which every careless tourist believes to be exactly alike, are not of the same size. The southern tower is narrower than the other. It has been suggested that this

PIER AND ONE OF THE DOORS OF THE PORTE STE. ANNE.

may have been to give better access to the bishop's residence, which was on that side, but the hypothesis is unnecessary. The difference is sufficiently explained by the dislike for exact repetition, which is a characteristic of living work in the fine arts. There are also differences in the details, sufficiently visible to give reasons for preferring one of the towers to the other. MM. de Guilhermy and Viollet-le-Duc preferred the larger tower, that to the north, as being more ample in its details and better executed.

A detailed description of the sculpture on the west front would occupy many pages, and be unreadable. Of the three portals, that in the middle has the Last Judgment for the subject of its tympanum; that on the north side illustrates the life, death, and glorification of the Virgin; that on the south side is more confused. It is called the *portail St. Anne*, but is composed of fragments illustrating the lives of Saint Anne and the Virgin also. It is curious for the adaptation of transitional work (from Romanesque to Gothic) to a purely Gothic purpose. As the carvings already existed, it seems to have been thought right to employ them, but they would not fit the new fashion of the pointed arch; so the space between the two kinds of arch had to be dissimulated by filling it up with an enrichment in sculpture. Notwithstanding the great ability of the architect, we may be allowed to remark that he did not manage his *raccord* so cleverly as he might have done. The lower arch should have been effaced, and the space above it filled with angels. One objection applies even to the most perfect Gothic tympana of this kind; namely, the varying scales of the figures, which deprive the composition of unity.

One of the strong points in Notre Dame is the preservation of a few of her fine old doors. Those of the Virgin and Saint Anne have still their magnificent original iron-work of the twelfth and thirteenth centuries. The common people used to believe quite seriously that it was the Devil himself who had helped the smith in exchange for his soul, as mere unaided human skill was unequal to such a task. There was also a popular belief that an enchanter had shut the porte Ste. Anne so that it could not be opened,—the fact being simply that for a long time it was disused.

The reader must excuse me if I do not enter into details with reference to the north and south sides of Notre Dame. We have not space for a study of the subject, and it is not of any special interest except as regards the buttresses, which are very massive, and from which spring two arches to prop the walls, one reaching to the wall of the higher aisle, by passing over the roof of the lower aisle, and another clearing the roofs of *both* the aisles in two leaps, with a rest on the wall between, and then giving its support directly to the lofty walls of the clerestory itself. Another notable feature in the north and south fronts is the great rose-windows in the transepts, which, from their height, may be seen from a distance.

Now, let us pass into the interior. The first thing that strikes anybody conversant with architecture, after

LES TRIBUNES.

Notre Dame and the Sainte Chapelle. 69

the first strong impression of size and majesty, is that the columns of the nave are massive and Romanesque in character, and not so lofty, relatively to the height of the vault, as the columns at Westminster or Amiens, not to speak of the extraordinary ones at Bourges. There is, in fact, room for no less than five of these columns between the pavement and the apex of the vault. When Notre Dame was begun the Romanesque spirit was only just passing into the Gothic spirit, so that the church is not quite completely Gothic as yet, though very nearly so. Its double aisles are a remarkable feature, of great value in giving mysterious distances with many intersections of the vaults. They run entirely round the building, and have allowed the architect the means of creating a great gallery above the inner aisle (which is wider than the external one); a gallery of much value in a cathedral where magnificent royal ceremonies were expected to take place. This gallery is always called *Les Tribunes* by French writers. The view we give is taken on the south side of the cathedral; and the reason why it seems to come to a sudden termination is because the transept occurs there. With the exception of the interruption caused by the transepts, this gallery goes round the entire edifice, and has four staircases of its own.[1] Not only is it very useful on great occasions, but it adds immensely to the elegance of the whole church, and looks all the more delicate and airy because it is lighted from the exterior.

[1] It also turns aside into the transepts to the extent of two large bays.

In most of the French cathedrals the *pourtour du chœur*, or aisle between the apse and the chapels, excels all other portions of the church in the variety of its perspective and in the delightful changes occurring at every step as the visitor slowly advances. When he walks down the middle of a straight nave between parallel rows of columns, he may be impressed by the grandeur that surrounds him, but he always knows what to expect. In the *pourtour* there is the new element of the unforeseen. He sees first one part of a chapel and then another; he loses one beautiful and intricate composition of columns, vaults, and windows, only to exchange it for another not less beautiful; and so attractive is the desire for what is coming, so keen the regret for what is left behind, that it is almost equally difficult to stay in one place or to leave it. This, at least, is what I have always felt in the few great *pourtours* which are comparable to that of Notre Dame. This one, in particular, has the additional intricacy of its double aisle, and now that it is enriched with painted glass and mural illumination the effect is at the same time more splendid and more mysterious than in the chilly eighteenth century.

This brings one to speak of the restorations which have been carried out at Notre Dame in our own day. Nothing is easier than to condemn such restorations absolutely; but those who do so cannot surely realize what was the state of such edifices as Notre Dame before the modern restorer dealt with them. It should be

THE "POURTOUR."

remembered that no age but our own ever had the slightest respect for the work of any preceding age, that we are the first human beings who ever valued the architectural work of our ancestors, the first who were ever pained by injury done to the work of another time, the first who ever understood that unity of design might be one of the merits of a building. Instances of injury done to great edifices *before* the modern restorer came are infinitely numerous; but I must here confine myself to Notre Dame. First, let us rapidly survey the exterior.

In the west front the row of statues called the Kings had been all cast down at the Revolution. Were the niches to be left empty? Certainly the original architect never intended them to be empty; his intention was that there should be statues, and the modern restorer fulfilled that intention, so far by putting statues there. The subjects are supposed to have been the Kings of Judah, and as the real faces of those kings have not come down to posterity in portraits, the present set of statues are as much likenesses as their predecessors. The important point was to have statues in keeping with the character of the building; and this was done as far as possible by copying such fragments of the old statues as could be found, and by imitating others in cathedrals of the same date. The restorer could not have done less, and it is not easy to see how he could have done more. Now let us pass to the central doorway. Among the lights of the eighteenth century was a famous architect called Soufflot, who fancied that he could improve upon Gothic ideas, and who, unfortunately, had the power to alter as well as to criticise. So he removed the pier between the doors, with the statue of Christ, and made a wide pointed arch in the middle of the tympanum, cutting into its elaborate sculpture as coolly as if it had been a common brick wall. Then he put classical columns, with modern doors, and was perfectly satisfied with his improvement. Could Lassus and Viollet-le Duc, the restorers, leave this incongruous absurdity untouched? Clearly not. They had the pier replaced, and they got an able sculptor, Dechaume, to carve a Christ for it, which he did after careful study of the statues at Amiens and Reims. The tympanum was restored as far as possible, and Soufflot's Renaissance doors were replaced by others more in keeping with those of the Virgin and Saint Anne. Surely, in this case also, it would hardly have been possible to do less. Other details might be dwelt upon if we had space; but let us consider a little what was the condition of the north and south sides. Let us hear Viollet-le-Duc's account of the state in which he found them: —

"They (the architects of the eighteenth and nineteenth centuries) had altered in the most deplorable manner the architecture of the sides of the nave. One might say that this portion of the edifice had been, as it were, *planed*. One after another the architects had suppressed the advancing parts of the buttresses between the chapels, the gables, the friezes, the balustrades, — in one word, the entire ornamentation of these same chapels, the pinnacles which decorated the tops of the buttresses with the statues that accompanied them and their flowering spires, the picturesque gargoyles which rendered the service of throwing the rain-water to a distance from the walls."

Were the restorers to leave the sides of the cathedral in this naked condition, or were they to attempt to adorn them again as nearly as possible according to the first intention of the builders? They decided to make the attempt, and they felt authorized to do so because they knew more about Gothic architecture, and had more love for it, than any other architects since the Renaissance. At the intersection of the roof there had been a light spire in Gothic times, — light, I mean, in appearance, made of oak, covered with lead. This spire was pulled down in 1793. Was its place to be left vacant? Certainly there was no inability to erect an elegant new spire, as the one now existing clearly proves. The architect Soufflot, who spoiled the great doorway, had built a vestry on the south side of the cathedral in a style which the reader may imagine. Part of it remained to our own day, but this was removed, and a new one erected by Viollet-le-Duc in thirteenth-century Gothic. There are two objections to this building: it looks rather too pretty and too intentionally contrived for the picturesque, and its newness is still out of keeping, but it does no harm whatever to Notre Dame. It would be difficult to suggest anything better.

Now, with regard to the interior. Here the ignorance and bad taste of the ages in which Gothic architecture was not understood had full play for several generations. The choir of a church is the part most richly furnished and decorated. In the Middle Ages the choir of Notre Dame was completely furnished with all the elaborate works of art which the feeling of the time held to be necessary in a great religious edifice; and down to the close of the seventeenth century these things were still in existence. There were the old stalls of the fourteenth century; there was a magnificent carved screen in open stonework going all round the choir; there was the high altar, with its columns of brass, its shrine of silver-gilt, its winged angels. All these things disappeared to make way for costly Renaissance decorations, which have been respected as far as possible by the modern restorers. In 1741 the Chapter gave orders for the removal of the splendid stained glass which filled the windows of the nave and choir; and a man called Pierre Levieil was ordered to replace them with common glass ornamented with a border of fleurs-de-lis. Levieil set about his work honestly and innocently, believing that it was quite proper to destroy what future ages could never replace, and he has left in writing some record of his doings. Regret for all the magnificence thus lost forever is happily tempered by rejoicing, as it most fortunately happened that the barbarians let alone the great rose-windows of the transepts and the west front. Modern art has endeavored in some measure to replace what was destroyed, being clearly authorized to do so by the intention of the original builders, who counted upon the effect of colored glass in tempering the excess of light. Viollet-le-Duc went a little further in one detail, for he took the opportunity of

ROYAL THANKSGIVING IN NOTRE DAME, 1782.

opening new rose-windows above the tribunes, near the transepts and choir, to recall the original arrangement by which such windows existed over the arches of the tribunes. This adds to the interest and peculiarity of the building, and has an historical reference.

All that remains to be said about the restoration is that the architects found Notre Dame entirely covered internally with thick coats of colored washes, which they removed for two reasons, — firstly, because they were hideous; and, secondly, because they prevented the masons from examining the condition of the stonework and making the necessary repairs.

The degree to which Gothic architecture was appreciated in the eighteenth century may be judged of by the fact that when the old painted glass was removed, the nave was turned into a picture-gallery, so as to hide every one of the arches, — as if there could be anything more necessary than its arches to the effect of a Gothic church! The pictures are now, happily, removed. Good or bad, they were equally out of place. Pictures, other than mural paintings of a severely conventional kind, always are out of place in spacious Gothic interiors.

The origin of the Sainte Chapelle is probably known already to most of my readers. It is nothing more than a large stone shrine to contain relics. Nothing could exceed the joy of Saint Louis when he believed himself to have become the possessor of the real crown of thorns and a large piece of the true cross. He bought them at a very high price from the Emperor of

Constantinople,[1] and held them in such reverence that he and his brother, the Count of Artois, carried them in their receptacle on their shoulders (probably as a palanquin is carried), walking barefooted through the streets of Sens and Paris: such was the thoroughness of the King's faith and his humility towards the objects of his veneration.

These feelings led Saint Louis to give orders for the erection of a chapel in which the relics were to be preserved, and he commanded Peter of Montereau to build it, which Peter did very speedily, as the King laid the first stone in 1245, and the edifice was consecrated in April, 1248. There are two chapels, a low one on the ground-floor and a lofty one above it; so both were consecrated simultaneously by different prelates, the upper one being dedicated to the Holy Crown and the Holy Cross, the other to the Virgin Mary.

Considering the rapidity of the work done, it is remarkable that it should be, as it is, of exceptionally excellent quality considered simply with reference to handicraft and to the materials employed. The stone is all hard and carefully selected, while each course is fixed with clamp-irons imbedded in lead, and the fitting of the stones, according to Viollet-le-Duc, is "*d'une précision rare.*"

[1] Some say that the crown of thorns was purchased from John of Brienne, the Emperor, and the piece of the true cross from Baldwin II., his successor; others say that both were purchased from Baldwin II. The cost to Saint Louis, including the reliquaries, is said to have been two millions of livres. So far as the King's happiness was concerned, the money could not have been better spent.

Notre Dame and the Sainte Chapelle. 77

Like Notre Dame the Sainte Chapelle has undergone thorough and careful restoration in the present century. For those who blame such restorations indiscriminately I will give a short description of the state of the building when it was placed in the restorer's hands. It had been despoiled at the Revolution and was used as a magazine for law-papers. The spire had been totally destroyed, the roof was in bad repair, sculpture injured or removed, the internal decoration mostly effaced, the stained glass removed from the lower part of the windows to a height of three feet, and the rest patched with fragments regardless of subject. The chapel was an unvalued survival of the past, falling rapidly into complete decay, and surrounded by the modern buildings of the law courts, so its isolation made total destruction probable. There had been a time when the Sainte Chapelle had been in more congenial company. The delightfully fanciful and picturesque old Cour des Comptes had been built under Louis XII. (1504), on the southwest side, and there was the great Gothic Cour de Mai, and, finally, the Great Hall on the north. Not only that, but there was the Trésor des Chartes, attached to the south side of the Sainte Chapelle, itself a treasure, almost a miniature of the glorious chapel, with its own little apse, and windows, and high-pitched roof. All these treasures of architecture were gone forever, replaced by dull, prosaic building; the Sainte Chapelle served no purpose that any dry attic would not have served equally well, and there seemed to be no reason why it should not be destroyed like the rest. The decision was to restore it, and give it a special destination as the place where the lawyers might hear the mass of the Holy Ghost. The work was done thoroughly and carefully by learned and accomplished men. M. Lassus designed a new spire,[1] an exquisitely beautiful work of art, much more elegant than its predecessor, as the reader may judge in some degree by comparing the etching with the woodcut.[2] Still, to appreciate the new spire properly, one needs an architectural drawing on a large scale, like that in the monograph by Guilhermy. It is of oak, covered with lead, with two open arcades. There are pinnacles between the gables of the upper arcade, and on these pinnacles are eight angels with high, folded wings and trumpets. Near the roof are figures of the twelve apostles. All along the roof-ridge runs an open crest-work, and at the point over the apse stands an angel with a cross. All these things, judiciously enlivened by gilding, with the present high pitch of the roof, add greatly to the poetical impression, especially when seen in brilliant sunshine against an azure sky.

Thanks to the restorers, the interior of the chapel once more produces the effect of harmonious splendor which belonged to it in the days of Saint Louis. Of all

---

[1] The spire by Lassus is the fourth. The first, by Pierre de Montereau, became unsafe from old age; the second was burnt in 1630; the third was destroyed in the Great Revolution.

[2] The woodcut is from a picture now at Versailles, painted by an artist named Martin in 1705. It is possible that he may have stunted the spire a little to get it into his canvas; he certainly has depressed the roof, unless the roof then existing fell considerably short of the original pitch.

THE OLD COURT OF ACCOUNTS AND THE SAINTE CHAPELLE.

Notre Dame and the Sainte Chapelle. 79

the Gothic edifices I have ever visited, this one seems to me most pre-eminently a visible poem. It is hardly of this world, it hardly belongs to the dull realities of life. Most buildings are successful only in parts, so that we say to ourselves, "Ah, if all had been equal to that!" or else we meet with some shocking incongruity that spoils everything; but here the motive, which is that of perfect splendor, is maintained without flaw or failure anywhere. The architect made his windows as large and lofty as he could (there is hardly any wall, its work is done by the buttresses); and he took care that the stonework should be as light and elegant as possible, after which he filled it with a vast jewelry of painted glass. Every inch of wall is illuminated like a missal, and so delicately that some of the illuminations are repeated of the real size in Guilhermy's monograph. When we become somewhat accustomed to the universal splendor (which from the subdued light is by no means crude or painful), we begin to perceive that the windows are full of little pictorial compositions; and if we have time to examine them, there is occupation for us, as the windows contain more than a thousand of these pictures. Thanks to the care of M. Guilhermy, they have been set in order again. The most interesting among them, for us, on account of the authenticity of the historical details, is the window which illustrates the translation of the relics. Here we have the men of the time of Saint Louis on land and sea. In the other windows the Old and New Testaments are illustrated. Genesis takes ninety-one compositions, Exodus a hundred and twenty-one, and so on, each window having its own history.[1]

There are four broad windows in each side, though from the exterior two of these look slightly narrower because they are somewhat masked by the west turrets. The apse is lighted by five narrower windows, and there are two, the narrowest of all, which separate the apse from the nave.

In the time of Henri II. a very mistaken project was carried into execution. A marble screen, with altars set up against it, was built across the body of the chapel so as to divide it, up to a certain height, into two parts. Happily, this exists no longer.

The original intention of Louis IX. when he built the Sainte Chapelle was that the upper chapel should be reserved for the sovereign and the royal house, while the lower one was for the officers of inferior degree. The King's chapel was on a level with his apartments in the palace, so that he walked to it without using stairs. The lower chapel has now been completely decorated like the upper one, on the principles of illumination. It is beautiful, but comparatively heavy and crypt-like, and the decoration looks more crude, perhaps

[1] The only thing in the Sainte Chapelle which can be considered in any degree incongruous with the unity of the first design is the rose-window at the west end, which was erected by Charles VIII. near the close of the fifteenth century. The flamboyant tracery is of a restless character, all in very strong curves, and the glass is quite different from the gorgeous jewel-mosaics of the time of Saint Louis. The subjects are all from the Apocalypse. However, this window inflicts little injury on the general effect of the chapel, as the visitor is under it when he enters, and it is isolated from the rest. In service time everybody has his back to it.

SAINT LOUIS IN THE SAINTE CHAPELLE.

because the vault is so much lower and nearer the eye. A curious detail may be mentioned in connection with the religious services in the Sainte Chapelle. They were of a sumptuous description, as the "treasurer," who was the chief priest, wore the mitre and ring, had pontifical rank, and was subject only to the Pope. He was assisted in the services by one chanter, twelve canons, nineteen chaplains, and thirteen clerks. When Saint Louis dwelt in his royal house close by and came to the Sainte Chapelle, the place must have presented such a concentration of mediaeval splendor as was never seen elsewhere in such narrow limits. His enthusiasm may seem superstitious to us, but he endeavored earnestly to make himself a perfect king according to the lights of his time, so that his splendid chapel is associated with the memory of a human soul as sound and honest as its handicrafts, as beautiful as its art.

V.

THE TUILERIES AND THE LUXEMBOURG.

SOME readers may ask why the Tuileries should be a subject for a chapter in a work on Paris, when the palace is a thing of the past, and the last stones of it have been carted away from its empty site.

To this objection there are two replies. The first is, that the historical importance of the palace will make some mention of it inevitable in every work on Paris for ages yet to come; because, if the stones are no longer there, the site must ever remain. The second answer is of a more positive and practical nature, making no appeal to feelings with reference to past history, which exist powerfully enough in some minds but are entirely absent from others. The Palace of the Tuileries has always been held to include the two blocks of buildings at the northern and southern extremities, called the *Pavillon de Marsan* and the *Pavillon de Flore;* and by some authorities the lines of building running eastward from these pavilions are held to belong to the Tuileries, as far as the pavilions de Rohan and Lesdiguières. Now all this exists at the present day, after much restoration, even after much reconstruction; and is still an architectural feature of Paris too important to be omitted.

Many readers of these pages will remember the Tuileries as they appeared in the time of Napoleon III. In those days the main body of the palace was a very thin and very long line of building, which extended from the Rue de Rivoli on the north to the bank of the Seine on the south; and included nine masses, each with its own roof. In the middle stood the Pavillon de l'Horloge, and at the two extremities, as I have just had occasion to observe, the pavilions Marsan and Flore. The remaining six masses of building were distributed symmetrically, three on each side the Pavillon de l'Horloge, but each pair of them differed greatly from the others.

The first impression produced by the Tuileries on a foreigner who knew nothing about its architectural history was that "it was a vast and venerable pile": —

> "Huge halls, long galleries, spacious chambers, joined
> By no quite lawful marriage of the arts,
> Might shock a connoisseur; but when combined,
> Formed a whole which, irregular in parts,
> Yet left a grand impression on the mind."

I remember that first "grand impression" well, and can easily recover it even now. The great length of the garden front gave a magnificent effect of perspective, ending admirably with the towering pavilions, and divided by the central pavilion and the range of different roofs which rose one behind another like mountains. The color was a fine warm gray, turned to a golden gray by the effulgence of sunset, when the long range of windows glistened in the evening light. It is said that on a certain day in the year when the sun was to be seen exactly within the great, far-away arch of triumph, the last of the French kings would come out on the balcony of the great central pavilion and watch the rare and magnificent spectacle. It is not very long since then, in mere numbering of years; and there are people still living who have seen the King on the royal balcony, yet it belongs even now as much to the past as the princely life at Nineveh. The last King lies, nearly forgotten, in the mausoleum on the top of the hill at Dreux, wisely chosen far from the capital, that the House of Orleans might rest in final peace; and where the long, picturesque old palace stood is a great gap of empty air.

The destruction of the Tuileries by the Communards was a lamentable event from the point of view of the historian and the archaeologist, but artistically the loss is not great. If the Empire had lasted, the palace would have been destroyed by architects, as the total reconstruction of it had been decided upon long before. In spite of the immense sums which at different times had been spent in making it habitable, it still remained one of the most inconvenient houses in the world. The extreme (relative) narrowness of it made communication troublesome and long, while there was a great want of proper corridors; and, in short, the structure was only the result of adding and mending, not the realization of a logical and orderly plan. I cannot say whether the projects for the new palace had ever been elaborated in the shape of finished drawings; if they were, it was

The Tuileries and the Luxembourg. 85

thought judicious not to show them to the public; but long before the fall of the Empire I was told, by one who knew the imperial intentions, that the old palace of the Tuileries was condemned. The first step was taken by pulling down the Pavillon de Flore, and when the new one was erected in its place, a short piece of new work, equally magnificent, was carried northward and stopped abruptly, to accustom the public to the idea of its ultimate continuation. At the same time it does not appear that Louis Napoleon contemplated the immediate rebuilding of the Tuileries, as he arranged a very beautiful and costly suite of private apartments for the Empress within the shell of the old palace.

Hardly any old country-house in England has been built in such a haphazard fashion. The first architect no more thought of uniting the Tuileries to the Louvre than the builder of Buckingham Palace thought of joining it to the Horse Guards; and yet this notion ultimately governed everything, entirely depriving the Tuileries of completeness and independence, and making it only part of a colossal whole, which, from the artistic point of view, was simply a colossal error.

The history of it begins in the year 1564, when Catherine de Medicis conceived the idea of having a palace to herself near the Louvre, yet independent, in which she might be near enough to her son Charles IX. to have influence over him. She wanted it to be a country-house, or what we should call suburban, just well without the walls of Paris. Here the reader must be reminded that in 1564 the wall of Paris was no longer that of Philippe-Auguste, which went through the present square of the Louvre, but that of Charles V., which went through what is now the Place du Carrousel, just to the east of the Salle des États, or a little to the west of the pavilions de Rohan and Lesdiguières. It was a fine strong wall, with square towers, and a round tower at the corner near the Seine, called the *Tour du Bois*, which remained long afterwards, and is a familiar object in old prints.

There is this very curious coincidence in the first construction of the palaces of the Louvre and the Tuileries. Each of them, in the beginning, was intended to be just outside the wall of Paris, the Louvre being west of the wall of Philippe-Auguste, the Tuileries west of Charles V.'s wall. The difference in the style of architecture adopted marks the difference between the temper of Gothic and Renaissance times. Philippe-Auguste built the Louvre as a strong Gothic fortress; Catherine de Medicis, with ideas imported from Florence, wanted a Renaissance palace of graceful architecture where she might dwell in elegance and comfort. She got her elegant dwelling, but had not much comfort there, as it happened.

And now, from an artistic point of view, comes the saddest part of the whole story. Catherine had a man of taste and even genius at her orders, the great architect Philibert Delorme, and he had a plan for a palace of moderate dimensions but of perfectly rational conception, — such a palace as would have been a really complete work of art, and a great ornament to Paris in our own day, had it been preserved so long. Catherine appreciated and employed him; but she was short of funds, and he unluckily only lived a few years, so that his complete plan could not be carried out in his lifetime, which would have settled everything.

As the name of Philibert Delorme is so closely connected with the origin of the palace, there is a common popular belief that at least the central pavilion and the wings nearest it were built by him, as we knew them, and such is the power of fame that they were often admired on the strength of his reputation. If his shade could have revisited the garden, and seen the front in the time of Louis Napoleon, he would probably have found more pain than pleasure in the knowledge that his name was connected with it at all. The whole of his work, even including the central pavilion, was altered by subsequent architects till the beauty and grace of it were effectually taken away. Delorme's building consisted simply of a ground-floor and an upper story which was lighted by beautiful dormer windows, with rich stone panels inserted between them. Above these rose a well-pitched roof, and care, of course, was bestowed upon ·the chimneys. But the most important feature in Delorme's design was the pavilion (he only lived to erect one, in the centre of his front). The basis of this pavilion was a strong square mass two stories high, with a central doorway between two pairs of columns, and a window above it, also between two pairs of columns. The whole square mass was surrounded by a balustrade at the top, and there was a large round dome standing upon an elegant arcade and accompanied by four small domes, occupying the angles of the square mass beneath. These satellites were supported on arches like the great dome, and on the top of the great dome was a lantern, also on little arches. The windows in the front were set in pairs near the pavilion and at the extremities, but between these pairs were three single windows;[1] the composition, as a whole, was extremely elegant, and, though quite palatial and fit for a queen, it was neither cumbersome nor pretentious. If the architect had lived, and if the queen had been richer, they would have completed a quadrangle measuring about 270 mètres by 168 in that manner, but with corner pavilions, one of which was erected by Jean Bullant on the south side after Delorme's death, which occurred in 1570, after he had worked eight years for Catherine de Medicis.

As the quadrangle was never completed, only one side of it having been built, the palace was found to be too small in later reigns, and so it was increased in length and in height, as I shall have to explain shortly, and Delorme's work was spoiled by heavy superposition. He had chosen the Ionic order as more feminine than the Doric, because the palace was for a lady. He himself gives this reason, the Ionic having been employed for the Temples of Goddesses. At the same time he gave the palace an air of elegance of which it was afterwards deprived.

It is remarkable that Catherine hardly used the

[1] This description is from what is now the Place du Carrousel.

The Tuileries and the Luxembourg.

Tuileries. It appears to be certain that she only went there as people go to a summer-house, for a few hours at a time, or, at most, for a very short stay, and that the palace was not even furnished, as the officers of her household sent on each occasion the furniture that she required, and had it removed when she was gone. The architectural works were completely abandoned by Catherine in 1572; either she was tired of her hobby, or there may be some truth in the commonly repeated tradition that she was frightened away from the Tuileries by the prediction of a fortune-teller.[1]

Some readers will remember the large space behind the Tuileries, between the palace and the railings across the Place du Carrousel. In recent times the space was nothing but an arid desert of sand, very useful for reviewing troops, but as monotonous as a barrack-yard. In the early days of the palace this was occupied by a beautiful garden, and even before the building of the palace was begun a fine garden, in the formal taste of the time, had been made to the west, on the ground occupied by the present garden of the Tuileries. There were six great straight walks going from end to end, and these were crossed by eight narrower walks at right angles; the beds were consequently all rectangular, and even within the beds the same rectangular

[1] The story is in the guide-books, so it is scarcely necessary to repeat it; but to save the reader the trouble of a reference I may say that the fortune-teller tried to be agreeable to her Majesty by predicting for her a quiet end "near St. Germain," as the Tuileries was in that parish. Catherine avoided the palace afterwards to prolong her chances of life, yet died near St. Germain after all, as the priest who attended her bore that name.

system was carried out in the subdivisions. At a later period, while the stone borders of the beds were preserved, there was a violent reaction against angles inside them, and the most intemperately curved flourishes were substituted. I have no doubt that this intemperance in curvature was the direct consequence of the straight-line system which had created a great hunger for curves. In Catherine's original garden there was not a single curve of any kind except the semicircle of the echo. With regard to the general principle of the formal French garden, it may be defended as a suitable accompaniment to symmetrical architecture. Such gardens, when of great size, are wearisome in the extreme; but a small one is valuable close to a building, as a sort of extension of the building itself upon the ground.

The new palace of the Tuileries had been so much neglected that when Henri IV. came to it he found it already nearly ruinous. He was one of the great building sovereigns; the constructive instinct was strong in him from the beginning, so of course the unfinished condition of the Tuileries excited him to architectural effort. Unfortunately for the future artistic consistency of these great palatial buildings, he conceived the idea of uniting the Tuileries to the Louvre by a long gallery on the river-side, which of course involved from the first the necessity of a corresponding line of building on the north, along what is now the Rue de Rivoli. The enterprise was so prodigious that nine sovereigns reigned over France before it was completed; and no sooner had it been finished by Louis Napoleon than the incongruousness of old and new made him decide to build the Tuileries over again. If Henri IV. had simply confined himself to carrying out the first intentions of Philibert Delorme by building the whole of that architect's projected quadrangle, the result would have been charming. What he actually did spoiled the Tuileries completely; he built the Pavillon de Flore, which, by its great mass, made Delorme's dome too small for its central position, and the heavy architecture of the long gallery, with its big pilasters from top to bottom, set an evil example for future work on the Tuileries. It is believed that Henri IV. built the long gallery for reasons of prudence, and that he desired to plan for himself a way of retreat in case of a popular attack on his palace of the Louvre. The reader is asked to remember that the Tuileries was still out of Paris, and that the wall existed still except where it was pierced by the new gallery. Henri had a private garden between the Tuileries and the city wall, and special precautions were taken to secure the completeness of its privacy.

It is an interesting fact that from the beginning the great gallery was used for works of art, while it served as a means of communication; and it is also a remarkable proof of the interest taken by Henri IV. in the arts, that he lent the extensive series of rooms on the ground-floor to workers in painting, engraving, tapestry, and sculpture. These rooms appear indeed to have been employed as schools of art; and French writers believe them to have constituted at that time a sort of *conservatoire des arts et métiers*, — a free school of fine and industrial art.

When Henri IV. had done his work the edifice must have presented a strikingly awkward and unfinished appearance. Fastened on one corner of the quadrangular Louvre was a mass of building going down to the quay, and from this the long gallery went to the Pavillon de Flore; the length of it not having been determined by any architectural consideration whatever, but simply by the distance which happened to exist between two different palaces. At the west end the appearance must have been most unsatisfactory. There was the big Pavillon de Flore, and a mass of building to connect it with the poor little palace of the Tuileries; and on the other side there was nothing. Between the Tuileries and the Louvre lay a confusion of garden, ditch, wall, and various habitations.

Henri IV. was able to walk under cover from one palace to the other in the last year of his life, but the device for escaping from the city did not save him from assassination. After him Louis XIII. went on with the work; but the great builder was Louis XIV., who was displeased with the one-sided appearance of the palace, and also with the extreme irregularity of the roofs. By that time the ditches and wall of Charles V. had been removed, and the east garden (called the *Jardin de Mademoiselle*) had been made into a desert; so on the 5th of June, 1662, the King held a great equestrian festival in the space between the Tuileries and the Louvre (but nearer to the Tuileries), from which that piece of ground has been called ever since then the *Place du Carrousel.* The festival was a mixture of costume cavalcades and games; the King himself took an active part in it, and so did the flower of French nobility. The minute accounts left by eyewitnesses make it certain that the scene was one of extraordinary splendor; but the architectural background was so incomplete, that perhaps the King's resolution to take up the work may date from that very day. Nothing could be done to save the Tuileries of Philibert Delorme. A great northern pavilion, the Pavillon de Marsan, was erected to make a northern angle answering to the southern Pavillon de Flore; and it was joined to the other buildings, but these were so disproportioned that it was thought necessary to raise some of them by adding another story (or more), and to bring the front more nearly to a level by building across its cavities. The central pavilion was raised a story, and a heavy dome with angular corners was substituted for the elegant round dome of the first architect. This was the Pavillon de l'Horloge, that we remember.

I have said that the Tuileries consisted of nine masses of building. It may be convenient to remember that the architect, Philibert Delorme, only completed three of these, — the central pavilion and two wings; Jean Bullant added a pavilion to the south. The architects of Henri IV. added two masses still farther to the south; those of Louis XIV. added three to the north, so that in his time the nine ultimately attained were already complete. It is difficult to see how his architects, Le Vau and d'Orbay, could have dealt effectively in any other way with the difficult problem before them, unless they had completely demolished the old palace. The real blunder was not committed by them, but by Henri IV. and his architects, Métezeau and Du Cerceau, when they made the work of Louis XIV. an inevitable necessity of the future.

We have clear evidence that in the time of Louis XIV. it was already intended to build the long northern side of the great square. An engraving by Israel Sylvestre, representing the famous equestrian festival, anticipates the future by showing the Pavillon de Marsan as already erected; and not only that, but he even shows the beginning of what was afterwards done by Napoleon I. to unite the Tuileries to the Louvre.

The Great Napoleon was not quite so passionately fond of building as Napoleon III., but he liked to leave his mark on Paris, and his military love of order and completeness was vexed by the confusion behind the Tuileries. Where the eastern garden once had been there were three spaces divided by hoardings, and also separated by hoardings from the rest of the Place du Carrousel, while there were a number of wooden booths within them, and a number of very ordinary houses just behind. It is surprising that preceding sovereigns should have tolerated such a state of things just behind their palace; and it is a remarkably apt

The Tuileries and the Luxembourg. 95

illustration of the wise old French proverb, "*Qui trop embrasse, mal étreint.*" The space included in the great scheme was so vast that it was never properly dealt with until our own time. Napoleon I. had two objects in view when he began his improvements: he first wished to keep people at some distance from the Palace for reasons of privacy and safety, and then he wanted a convenient place for small reviews of troops. He therefore cleared away all the hoardings and booths, and made an open gravelled space, which he separated from the Place du Carrousel with a railing. He also made it his business to clear away the houses and to build the north side according to the intentions of Louis XIV., in a plain, rather heavy style, with tall pilasters, suggested by the long gallery of the Louvre.

The work done by Louis-Philippe was considerable, but principally in the interior. The details of these changes would not greatly interest the reader, and would scarcely be intelligible without a plan. They included a new grand staircase, a new great saloon, and the improvement of the Galerie de Diane, with other alterations, which placed the floors of a long series of state apartments on the same level. These rooms in the aggregate were eight hundred feet long, and the bill for these improvements reached the handsome sum of £211,656.

Then came Louis Napoleon, who determined to complete the whole vast edifice of the united palaces. He had the builder's passion quite as strongly as either

Henri IV. or Louis XIV.; and during those years when nobody could resist his will, he indulged it to the uttermost. The greater part of his work belongs to the Louvre, as it lies east of the pavilions de Rohan and Lesdiguières, but he did much to the Tuileries of Henri IV. He pulled down the Pavillon de Flore, and rebuilt it, and he did the same for all that part of the long gallery that used to have long pilasters. In the execution of this important work every opportunity for improvement that was consistent with a respect for the original idea was seized upon with avidity. The long pilasters were abandoned, and the new work treated in stories, like part of the older Louvre, with much elegance of design and richness of sculptured detail. The Pavillon de Flore was in some respects more ornate than its predecessor, especially in the upper parts; and on the whole it was a more lively composition, with better contrasts of effective sculpture and plain wall surface. An unquestionable improvement was in the roofs, which were made rich enough in lead-work to accompany the sculptured ornaments of the walls. The tiresome length of monotonous gallery running eastward from the Pavillon de Flore was happily and intentionally broken by the large gateway called the *Guichets des Saints Pères*, by the twin pavilions of that gateway, and the masses of building on each side of them, which are loftier than the roof of the gallery. Besides this, the space comprised between the Pavillon de Flore and the Guichets is itself wisely interrupted by a minor pavilion rising

THE TUILERIES IN 1837.

The Tuileries and the Luxembourg. 97

above the cornice, though not above the roof. By these devices the great fault of the river front, inordinate length, is made less visible. As for perfection of detail, there has never been any epoch of French architecture in which the essentially national style was worked out with more thorough knowledge and skill than under Napoleon III.

It is a constant pleasure to examine such good workmanship closely, to see what a remarkably high level the decorative sculptors have attained when none of them disgrace the rest. Much as we admire Gothic architecture, we have to acknowledge that the modern work on the Tuileries is what Gothic sculptors could never have accomplished. The renewal of the art by the study of Greek antiquity was a necessary preparation for palatial work of this kind.

It is a pity (from our present point of view) that Louis Napoleon did not remain in power long enough to rebuild the Tuileries with the help of M. Lefuel, who erected the new Pavillon de Flore. The new palace would, no doubt, have been lofty and massive enough to hold its own against the new buildings of the Louvre; and the central pavilion, especially, would have been a stately and imposing work of great size and magnificent decoration. The intended imperial palace is, however, gone to the shadowy realm of the things that might have been. In the place it was to have occupied we have seen for some years a blackened ruin; certainly one of the most beautiful and interesting ruins that ever were, and so impressive by its combination of dire disaster with still visible traces of royal splendor that only a poet could describe it adequately. Meissonier has worked in it carefully, and his minutely faithful brush will preserve for posterity those fire-crumbled columns, those shattered walls on which were still to be seen strangely preserved spaces of gold and color, as in some ruin at Pompeii. Even the king's balcony was still there, and the sunset light, indifferent to human vicissitudes, refreshed its gilding in the summer evenings.

What the Republic has done since its establishment may be told in a few words. The fire had destroyed the Pavillon de Marsan and much of the line of building along the Rue de Rivoli. These have since been rebuilt, as magnificently as the new Pavillon de Flore and the new part of the great gallery on the water-side. There appears to be an intention of continuing the work in the same style as far as the Pavillon de Rohan, or perhaps of erecting some great hall to break the line, for the new work stops abruptly; and as the new building is much broader than the old, the walls can never meet. The architects of the new portion have avoided the heavy long pilasters of Napoleon I., and adopted the more elegant system of division in stories already so successfully carried out on the south. No decision has been arrived at yet (1885) with regard to the space occupied by the destroyed buildings of the Tuileries. All that is certain is that nothing will be joined to the pavilions of Marsan and Flore, as these pavilions are finished on three sides. The open space seems to call for a noble edifice of some kind, and it is probable that some public building will ultimately be erected there. If this is ever done, it will be highly desirable that it should be set further back towards the Louvre, so as to give to the two great pavilions the effect of advancing wings. This would do more than anything to relieve the great length and monotony of the garden front.

Through all their errors and experiments the architects of the Tuileries and Louvre have been developing a style of architecture which, in its ultimate stage, is really imposing and palatial. The great pavilions are very nearly related to towers, and their steep square roofs are like truncated spires; but the idea is so completely adapted to the needs of a palace that we forget its origin in mediaeval churches and fortresses. Such pavilions are useful and necessary in edifices where the lines of building are long. They serve as landmarks, and by their perspective they enable us to measure easily the scale of the whole edifice. The full maturity of this architecture has only been reached in the present generation. The new parts of the Tuileries are finer than the older work which they replace, — finer, not only as being more magnificent, but because, after so many experiments, the resources of that kind of art have come to be better understood. A contemporary French architect eminent enough to be employed on a national palace would naturally produce more elegant work than the old river-front of the long gallery or the alterations made under Louis XIV. The principles of this architecture having been settled, it has reached that mature stage when nothing remains to be done but to perfect the application of them in detail.

I have not had space to speak of the historical interest of the Tuileries, and can only do so now on the condition of extreme brevity. The palace was never very long or very closely connected with the history of the monarchy. It is not at all comparable to Windsor in that respect. Henri IV. liked it, Louis XIV. preferred Versailles, Louis XV. lived at the Tuileries in his minority. The chosen association of the palace with the sovereigns of France is very recent. Louis XVI. lived in it, and so did Charles X. and Louis-Philippe. The two Napoleons were fond of it, perhaps because it gave them a better appearance of sovereignty than a new residence could have done. The last inhabitant was the Empress Eugénie, as Regent, and her flight has a pathos surpassing the flights or last departures of other sovereigns, since we know that the palace was never again to be brightened by either royal or imperial splendor.

The parliamentary history of the Tuileries is important, as it has been not only a palace, but a parliament house. In old times the royal stable was to the north, close to what is now the Pavillon de Marsan, and in the present Rue de Rivoli. The exercising-ground was in a long, narrow enclosure, which occupied the ground of that street as far as the Rue de Castiglione; and at its western extremity there was a building called the *manège*, which served as a parliament house for the Assemblée Nationale, while Louis XVI. lived in his

THE LUXEMBOURG AS IT WAS BUILT.

The Tuileries and the Luxembourg.

apartments in the palace and rarely came out of them. In May, 1793, the Convention began to sit in a newly arranged parliament house within the walls of the palace itself, and for some time after that the palace included Government Offices of all kinds, so that the first rough-and-rude beginnings of popular government in France were carried on in the royal house itself. The reader may be reminded also that Napoleon's *coup d'état* of the 18th Brumaire took place within the Tuileries, where Parliament was then sitting. The most important events in the Tuileries have sometimes been simply the arrival of a courier with news, or its mere reception by the quiet-looking telegraphic wire. I was in Paris when that little wire brought to the Emperor's private cabinet in the Pavillon de Flore the terrible news about Maximilian. I stood with a friend and looked on the sunny outside of the great palace, and we said, "It is a dark day for Napoleon III." From that day everything went wrong with him till he was laid in the sarcophagus at Chiselhurst.

The Tuileries and the Luxembourg have this in common, that each was built by a queen, and that each of the queens was a·Medicis. Marie de Medicis began her palace in 1615. Unlike the elder edifice, it has preserved at least its original character, but in order to obtain more room in the interior the garden front has been replaced by a new one farther out; and though the original style of the building has been carefully imitated its proportions have been inevitably destroyed. Unluckily, too, the addition (begun in 1836 and finished in

1844) was of a nature to increase the only serious defect of the first design, which was the doubling of the southern pavilions. The first plan may be briefly described as follows: there was a quadrangle with one pavilion at each corner towards the street, but two pavilions at each corner (or very near it) towards the garden. The garden pavilions were so near each other as to lose the advantage of perspective and appear heavy. The enlargement carried out by M. de Gisors, Louis-Philippe's architect, consisted in constructing two new pavilions in the garden close to the four already existing, so that at the south end of the palace there are now six heavy pavilions, three on each side. The new ones were connected by a new front which gave great additional space inside for a library and senate-house; but the result externally was to make the heavy end of the palace look heavier still. Nevertheless, as the building had to be enlarged to receive the senate, it is very difficult to see how any equivalent increase of size could have been conveniently obtained with so little deviation from the first design. The garden front is practically the same, the interior of the quadrangle is untouched, at least so far as this alteration is concerned, so is the street front, and it is only the east and west sides which are lengthened without any alteration in their style.

The architecture of this palace is not at all comparable, so far as the one quality of elegance is concerned, with the most beautiful parts of the Louvre and the Tuileries, but it is serious and dignified, and almost in faultless taste in its own grave way. It would be difficult to find a more appropriate building for a senate-house. The situation is pleasant and easily accessible, while the great space of beautiful garden gives the palace a degree of quiet not always attainable in a great city, and which, we may suppose, ought to be favorable to legislative deliberations. It is thought more prudent, in France, not to have the two Chambers in one building; and it was principally for this reason that a recent proposition to rebuild the Tuileries, as a great parliament house for both Chambers, met with few if any adherents.

The garden of the Luxembourg is a precious breathing-space for that part of Paris, and is still of fine extent in spite of its mutilation at the south end, one of the very few attempts at economy made by the Imperial Government. It has a great population of statues, including many portrait-statues of famous Frenchwomen; but the charm of it in spring and summer is in the abundance of bright flowers, fresh well-watered grass, and graceful foliage. The reader must not expect from me any adequate description of a garden, as I greatly prefer wild nature to all gardens whatsoever; but if I were compelled to choose between the lawns and alleys of the Luxembourg and a dusty street pavement, I would bear with the artificiality of the horticulturists.[1]

[1] I have said nothing of the interior, which is inaccessible to the public, with the exception of the galleries, about which there is nothing in the slightest degree remarkable, except some of the pictures and statues which they contain, and which lie outside the scheme of these papers.

VI.
THE LOUVRE.

THE present writer once met, in Paris itself, with a very prosperous manufacturer from Yorkshire, who was not at all aware that there were any pictures in the Louvre. He considered it " a good, large building," but had never heard of its connection with the fine arts; and it is believed that he returned to his native county without having visited the interior.

This case, among visitors to Paris, is no doubt very exceptional, and there are even great numbers of people in the world who have never been to Paris, and are yet perfectly aware that the Louvre is a palace of the fine arts. For myself, so far as memory can go back into the hazy land of childhood, I can still recover the dim grandeur of the as yet unknown Louvre, a palace of colossal, fantastic architecture, like a dream, with endless halls filled with solemn, sombre pictures in heavy gilded frames. To see the reality was the longing of my youth, and when at last I found myself in that interminable gallery of Henri IV., it seemed as if the whole earth could not offer a delight so glorious.

Meanwhile — and in this I resembled nearly all other English tourists — I knew nothing of the noble castle

THE LOUVRE IN ITS TRANSITION STATE FROM GOTHIC TO RENAISSANCE.

which the present Louvre had replaced. It seemed to me that the building had been made entirely as a museum for works of art, chiefly pictures, and that nothing of any consequence had ever stood upon the ground it now occupied. Deeply interested in all remains of the Middle Ages that were to be seen in my native island, and passionately mediaevalist at heart (for all young people who care at all about the past are enthusiasts for some particular epoch), I little dreamed that one of the most romantic royal castles that ever existed once stood on the ground now occupied by chilly halls of antique sculpture. Such a castle, if its ruins yet rose on some lonely height by the Seine, would be visited by every tourist, and sketched by every landscape-painter; but as it had the misfortune to be enclosed within the walls of a very great city, where the past is effaced to make way for the present, as accounts are sponged from a slate, not a stone is left standing, and only the learned have measured its site or counted its lordly towers. Yet the time when they were new and perfect, with conical roofs and gilded vanes, is not exceedingly remote from us in the great past of history; and if they could have been simply left undemolished, even without repair, we should still have had an unrivalled example of the fortress-palace of the Middle Ages. The buildings formed an oblong court with round towers at the angles and in the middle of the sides, while nearly in the centre of the court stood a massive round keep, and to the south and east were well-defended gateways. All this was moated, and on the side towards the river were other walls and towers, the last of which maintained a threatened existence down to the seventeenth century.

The origin of the word *Louvre* is believed to be a Saxon word, *Leowar* or *Lower*, which meant a fortified camp. Littré, however, does not go so far as this, but contents himself with the base-Latin *lupara* or *lupera*, which was a subsequent creation as a latinized form of *louve*. Surely no two words could be more distinct than *louve* and *louvre*, while *lower* (pronounced, of course, by all French people as *lovver*) is a very near approximation to the name of the modern palace. Nor is there any reason to imagine a connection between the castle of Philippe-Auguste and a she-wolf, whereas, in its scheme of fortification, it bears a striking resemblance to a Frankish moated camp. In " Paris à travers les Ages" M. Fournier borrows a drawing of one of these camps from Viollet-le-Duc's "Dictionary of Architecture," and the resemblance of its plan to that of the Louvre Castle is most striking. It stands near a river, which defends one of its sides; it is moated just as the Louvre was; the central round tower is placed in the great enclosure precisely in the same position; the gateways are in the same places, and the principal part of the fortress is withdrawn somewhat from the river, with an extra defence towards the river-side, exactly as in the Louvre Castle. There seems, then, to be no reason for doubting that the name of the present picture-gallery is due to the early use of its site for military purposes.

Although nothing of the Louvre Castle is now visible from the exterior, there still exists a small remnant of it

THE LOUVRE, FROM THE SEINE. FROM A DRAWING BY H. TOUSSAINT.

DETAILS BY PIERRE LESCOT IN THE QUADRANGLE.

enclosed within the modern palatial buildings. There is a considerable piece of the old wall in the Salle des Cariatides, and even a small corkscrew staircase which belonged to the old castle.

The transformation of the castle into a palace began long before the present Renaissance palace was thought of. The first step was a consequence of the enclosure of the Louvre within the walls of Paris. Under Philippe-Auguste it had been outside, under Charles V. it was within the wall; and therefore, being no longer a fortress dependent on its own strength for resistance, it could be made more habitable without danger. Charles V. increased its height for the purpose of giving more room, and made great alterations in the arrangement of the apartments. Under that sovereign the Louvre still retained all the appearance of a feudal castle. The moat still surrounded it, and all the towers, including the great keep, were still in their places; but the general aspect was richer and more elegant than before, the towers were loftier, the masses of building between them had become more spacious, and some new and graceful domestic architecture had been added within the courtyard. Lovers of books remember this epoch in the history of the Louvre in connection with the royal library which was established there. It is unnecessary to observe that even a royal library in the fourteenth century was but a small collection; and yet if that library of Charles V. could have been preserved to our own day, few collections would have been more valued by the curious. Some rooms in a particular tower were set apart for it, two rooms at first, and afterwards a third above them, the whole containing rather more than nine hundred volumes. The collection was afterwards increased, and amounted in 1410 to 1,125 volumes, many of which were afterwards lent or lost; and it is said that the Duke of Bedford carried off the remainder with him to England, after a sort of purchase, in 1429.

After being a splendid Gothic palace the old castle of the Louvre was almost entirely abandoned by the French sovereigns, and was employed as a prison and an arsenal. Then succeeded a long period of utter confusion, during which the new Renaissance palace was gradually coming into existence, while the remnants of the Gothic castle were devoured one after another, looking more and more miserable as less remained, till the wonder is that so late as Callot's time anything should have been preserved at all.

The appearance of Francis I. upon the scene is the doom of the old castle. With the help of an inventive and tasteful architect, Pierre Lescot, he began the Louvre that we know, — colossal in scale, magnificent, palatial, — utterly different in all ways from the domestic architecture of the great building sovereigns who preceded him; a building of which Philippe-Auguste and Charles V. could have had no conception whatever; a wonderful result of the study of antiquity, and of its influence coming to the French through the Italian mind.

What a strange revolution it is, how radical, how complete! The beautiful and picturesque French Gothic cast aside as barbarous, and, in its place, not at all a dull imitation of the antique,[1] but rather a new modern

[1] It is curious that Frenchmen in the time of Francis I. always spoke as if the new style were simply an imitation of the antique. They did not realize the fact that it was something more.

art having its roots far away in the past of Greece and Rome, and drinking nourishment from those distant sources. Imagine a French sovereign brought so completely under this new influence as not to care in the least for the beautiful Gothic art which had so delighted his ancestors! Charles V. had taken an honest pride in his Gothic towers, his tapestried halls, his comfortable wainscoted parlors, the round rooms where his books were kept; we know that he was proud of them because he showed the place himself to the Emperor. Had the old Louvre castle come down to our own times, it would have been restored in every detail with scrupulous accuracy, like Pierrefonds; and every mediaevalist in Europe would have visited it. Paris would have preserved it, as she now preserves the Hôtel de Cluny or the Sainte Chapelle. But Francis I. did not care about it in the least. Everything Gothic had gone completely out of fashion, and whatever he built was to be in the new Renaissance manner. He therefore deliberately began certain buildings at the Louvre which must, of necessity, either establish a permanent incongruity, or compel his successors to remove every fragment of the old castle. If any Parisian of those days yet held the Gothic times in affection, he must have foreseen regretfully the ultimate consequences of this new departure. "*Ceci tuera cela,*" he must have said to himself. Contemporary expressions of regret have come down to our own times; especially for the great tower, which was first demolished. After that the old castle seemed to take a new lease of existence. It was furbished up

THE CLASSICAL PAVILION AND THE OLD EASTERN TOWER.

The Louvre.

thoroughly to receive the Emperor Charles V. The scene of the well-known picture by Bonnington of the King and the Emperor visiting the Duchess d'Etampes was probably in the old Louvre.[1]

The new structure was begun in a very strange manner. The first part of it erected was a great classical pavilion, occupying the site of the southwest corner tower; and from this went a line of classical building as far as the Gothic southeastern tower, which was preserved. It is impossible to conceive an effect more incongruous than that of these huge new buildings introduced into an old Gothic castle of moderate dimensions.

Francis I. did little more than decide the fate of the old Louvre by introducing the new fashion. His successors went on with the work; and the progress of it may be followed, reign after reign, till the last visible fragment of the Gothic castle had been ruthlessly carted away. The northeastern and southeastern round towers are still to be seen in Israel Sylvestre's etchings done in the year 1650. It is very remarkable that the short building which connects the Louvre with the long gallery on the water-side, and which now contains the Galerie d'Apollon, should have been first erected, as well as a considerable portion of the long gallery itself, when the great square had as yet made no approach to completion. The scheme appears to have been from the beginning of the most confused kind. A liking for the water-side, and a consequent tendency to build in that direction, appear to have entirely overruled what-

[1] An etching from the picture by Flameng appeared in the "Portfolio" for January, 1873.

ever intention there may have been to carry out a decided plan. As soon as the erection of the Tuileries had been decided upon, the notion of a long gallery from one palace to the other began to fix itself in royal minds, and this long before the Louvre itself was finished. Charles IX. began the long gallery at his mother's instigation, and when Henri IV. finished it, neither the Tuileries nor the Louvre presented anything like a complete appearance. It is the strangest story! Image an English sovereign, too poor to complete either Buckingham or St. James's palace, spending vast sums in a line of building to connect them! The conduct of Catherine de Medicis is more wonderful still, for when neither the Tuileries, nor the Louvre, nor the connecting gallery, was finished, she began (with these three huge enterprises on hand) a new and most costly palace in a different part of Paris.

While the long gallery was slowly proceeded with, and the great new buildings had gone no farther than the western side of the great quadrangle, there was a confusion of buildings round about these great structures which it is surprising that a powerful sovereign could tolerate. The rulers of France, in the midst of the most gigantic plans, lived surrounded by eyesores. It has been supposed that Henri. IV. intended to clear the ground and embellish it with a garden, but he did not live long enough. Vast as is the Louvre that we know, it is as nothing in comparison with the prodigious scheme imagined by Richelieu and Louis XIII.; a scheme which, though never carried out, gave a very strong impulse to the works, and insured the completion of the present building, at least in a subsequent reign. It is probable that of all palace-building ever seriously imagined by a prince, the Louvre of Louis XIII. was the most colossal. If the palace contemplated by him had been carried out, it would have extended to the Rue St. Honoré, and included four great quadrangles of the same size as the present quadrangle, which, in its turn, is four times the size of the old castle of Philippe-Auguste. Nothing is more remarkable in the history of royal living than the great increase of scale that came in with the Renaissance. In the old Gothic times kings were contented with houses of moderate size, and with the exception of the great hall where the retainers assembled, the rooms were seldom very large; but no sooner had the Renaissance revolutionized men's ideas, than kings everywhere suddenly discovered that vastness was essential to their state. In France this new idea began with Francis I., and it is curious to observe how it worked out its full consummation. He began, as we have seen, with a spacious royal pavilion in the place of a narrow round tower. After him, the long gallery was conceived and executed. Then Louis XIII. imagined an immensity, which he only partially executed; finally, Louis XIV., still preoccupied by the same idea of hugeness, imagined another immensity, but this time outside of Paris, — at Versailles, — and executed it. Thus at length the new demon of the colossal got satisfied.

Happily for the Louvre, Louis XIV. interested himself in it before he engulfed his millions at Marly and

Versailles. While still quite young he felt urged to set to work by the provokingly incomplete appearance of the palace. Although Louis XIII. had demolished the last towers of the Louvre Castle, he had not done very much towards the completion of the palace. Only two sides of the quadrangle — the western and the southern — were as yet erected. Louis XIV. determined to build the two others, and as he had a clever and laborious architect at his disposal, the work advanced rapidly. We see Le Vau's work at the present day in the interior of the courtyard; but outside, especially towards the river, it has been modified or concealed. The story of this able architect, and his labors and tribulations, is one of the most pathetic in the history of the fine arts. It appears to be the doom of great architects, from the earliest times to our own, to be plagued by their employers, and compelled either to modify their plans or abandon them; but few have had to bear such mortifications as Le Vau. The reader no doubt remembers that eastern end of the Louvre where the great colonnade is. That was the beginning of his troubles. He had made his plans for that part of the outside, which, in his opinion, was of paramount importance, and had even begun its actual construction, when Colbert became superintendent of public works, and put a stop to it. Rival architects were appealed to for their opinion, and of course they all condemned Le Vau, who up to that time had been preferred to them. Not satisfied, however, with their propositions, or not feeling himself competent to decide among so many divergent pro-

THE INTERIOR OF THE QUADRANGLE. FROM A DRAWING BY H. TOUSSAINT.

fessional schemes, Colbert sent their drawings to Rome to have the opinion of the Italian architects of the day. In those days Italian architects were as firmly convinced that nobody but themselves knew anything about architecture, as are the French painters of the present day that English artists cannot have any knowledge of painting; so their decision might have been accurately foretold. They simply condemned everything that was sent to them, and said that the French sovereign stood in need of a real architect, who must of course be an Italian. Louis XIV. allowed himself to be dictated to by men who were supposed to be the leaders of Europe in architectural matters; and he engaged the famous Bernini, who came to Paris animated by such a sense of his own importance that he not only treated Le Vau and his plans as non-existent, but claimed the right to remodel the entire edifice without regard to the intentions of the earlier architects, Pierre Lescot and Le Mercier. Everything in Bernini's project was to be subordinate to stately architectural effects. The convenient arrangement of the interior was of no consequence to him, and it is said that he even failed to provide for the comfortable accommodation of the sovereign. Notwithstanding these very strong objections to Bernini, he seems to have imposed himself for a while so that works in stone and mortar were actually commenced under his superintendence. Bernini was treated like a prince, — paid, lodged, and served magnificently; but he did not produce a satisfactory impression, and many French influences united themselves against him, so on his departure to winter in Italy it came to be understood that he should not return; and he was consoled with a sum of three thousand louis d'or, and a life pension of twelve thousand livres for himself and twelve hundred for his son.

Then came a very strange thing in the history of the Louvre. Claude Perrault, a doctor of medicine and amateur architect, had elaborated a plan of his own for an east front, but had carefully refrained from putting it forward when the plans of the professional architects were sent to Italy, to be condemned by the national prejudice of the Italians. When Perrault's plan was shown to Louis XIV., the King had had enough of foreign opinion, and even of professional home opinion, and was in a humor to judge by himself. He had only two projects left to choose between, — that of Le Vau (modified and enriched) and the new one proposed by Perrault. Unfortunately for poor Le Vau there was a stateliness in Perrault's colonnade which pleased the pompous mind of the great King, so it was adopted with very little regard to suitableness. The final discomfiture of Bernini was most fortunate for the Louvre in one respect, — it saved the great quadrangle which Bernini wanted to spoil in various ways, especially by putting huge staircases in the four corners; but though the interior of the quadrangle was saved, it cannot be said that the adoption of Perrault's plan was by any means an unmixed benefit. The east front does not really belong to the edifice; it is merely stuck on, and when it was built the fatal discovery was made that it did not fit. Surely this cannot have been a *mistake*, in the common sense of the word, as a joiner makes a mistake of an inch in a piece of wood. Perrault's front was more than seventy feet too long for the building it was to be applied to. He must have known this. Most probably he was determined to have his fine long colonnade at all costs, and so deliberately exceeded the measurements at each end, regardless of the consequences, which were sufficiently serious. It became necessary to advance the river front farther towards the river. It was quite new. The architect who had built it, Le Vau, was still alive, yet the huge extravagance of building another, to mask it, had to be committed. This was the last drop of bitterness in the cup of sorrow served to Le Vau in his old age.

The consequence of Perrault's audacity is that the buildings on the south side of the quadrangle are much thicker than those on the other sides. It was not thought necessary to advance the north front in the same way, but the length of Perrault's colonnade made it necessary to build a projecting mass at the northeast corner. The external north front always seemed to have received less attention than the others, though now, in consequence of the much-frequented Rue de Rivoli, it is as much seen as the colonnade itself.

The colonnade has a great reputation, and is no doubt majestic and noble in its proportions, but it is wonderful how little it seems to belong to the building. This effect of being something separate is felt more strongly when we come out of the quadrangle by the east entrance, and then look back on Perrault's front. In all the alterations executed about the palaces nobody has ever touched that front; and, indeed, it is evidently one of those works that do not admit of change. Like all severely classical conceptions, it is an organic whole from which every diminution would be mutilation, and to which every addition would be an excrescence.

The western front of the Louvre remained extremely simple until the time of Napoleon III., when a feeble attempt was made to decorate it with some applied ornament, so that it might hold its own against the new buildings; and when this was found to be impossible it was masked by a new front of adequate magnificence. Until our own time this west front looked upon an accidental agglomeration of the commonest dwelling-houses, which filled what are now the Squares du Louvre and the Place du Carrousel. The completion of the great project, by which the Tuileries and the Louvre were to be united, has led to the clearance and embellishment of these spaces.

One of the greatest difficulties about the union of the two palaces was that they were neither parallel nor at right angles to each other. The degree of inclination is such that if a line drawn along the front of the Tuileries, and another along the west front of the Louvre, were both prolonged northward, they would meet within the walls of Paris near La Chapelle. Every architect who had studied the union of the palaces had proposed some means of hiding this defect. In 1810 no less

QUADRANGLE OF THE LOUVRE, WITH THE STATUE OF FRANCIS I., PLACED THERE IN 1855, AND SINCE REMOVED.

than forty-seven different projects were submitted to the Government. That of Percier and Fontaine was accepted, but never carried out. Those architects intended to hide the defect by carrying a line of building from north to south, straight across the middle of the Place du Carrousel. When Napoleon III. came into power he found Visconti in office as architect of the Louvre, and Visconti had another plan, which was executed. If the reader will refer to any recent map of Paris, he will understand Visconti's scheme at a glance. It consisted in the creation of a new court as wide as the inside of the old quadrangle, but longer, and open at the west end, in the direction of the Tuileries. Behind these massive lines of building are smaller enclosed courts to the north and south, the irregularity of which is only seen by the few who visit them. By this means it was hoped that the want of parallelism between the Tuileries and Louvre would be in a great measure concealed; but, unfortunately, the new buildings only made it more visible, by directing the eye towards the Tuileries in such a manner as to show plainly that the Pavillon de l'Horloge was not in the middle of the view. Again, it is easily seen from the Place du Carrousel that the new buildings do not occupy the same space on the north and the south sides. If, however, they are a failure as a means of hiding a defect, they have certain merits of their own. Considered in themselves, as examples of magnificent palatial architecture, they deserve little except praise; but in their relation to older buildings round the Place du Carrousel they were from the first objectionable, because their imposing size and rich ornamentation made everything else look thin, and low, and poor. To borrow a term employed by painters, the huge Visconti buildings simply "killed" the Tuileries. Their erection was the doom of the older palace by making a grander one a necessity of the future. The new pavilions Richelieu, Denon, Turgot, and Mollien, being very splendid in themselves and near together, made the Pavillon de l'Horloge of the Tuileries look miserable and lonely. Besides this, the massive lines of building that connected Visconti's pavilions, with their richly carved arcades surmounted by colossal statues, and their numerous groups of sculpture on the balustrades in front of the roof, made the long wing built (or begun) by Napoleon I. look fit for little else than a barrack-yard; and so we see it already replaced, in great part, by a much more magnificent structure, which will certainly join Visconti's buildings ultimately at the Pavillon de Rohan. It is narrated that Napoleon III., after gazing one day with a friend at the new buildings from a window in the Tuileries, turned away with a look of disappointment, and said, "If I listened to my own feelings I would begin the whole thing over again." There are limits, however, even to the extravagance of a Napoleon III.; and though he might easily have squandered as much in other and less visible ways, he could hardly indulge in such a public *repentir* as the reconstruction of his own Louvre.

The most obvious defect of Visconti's Louvre, considered in itself, is that the two great fronts which face

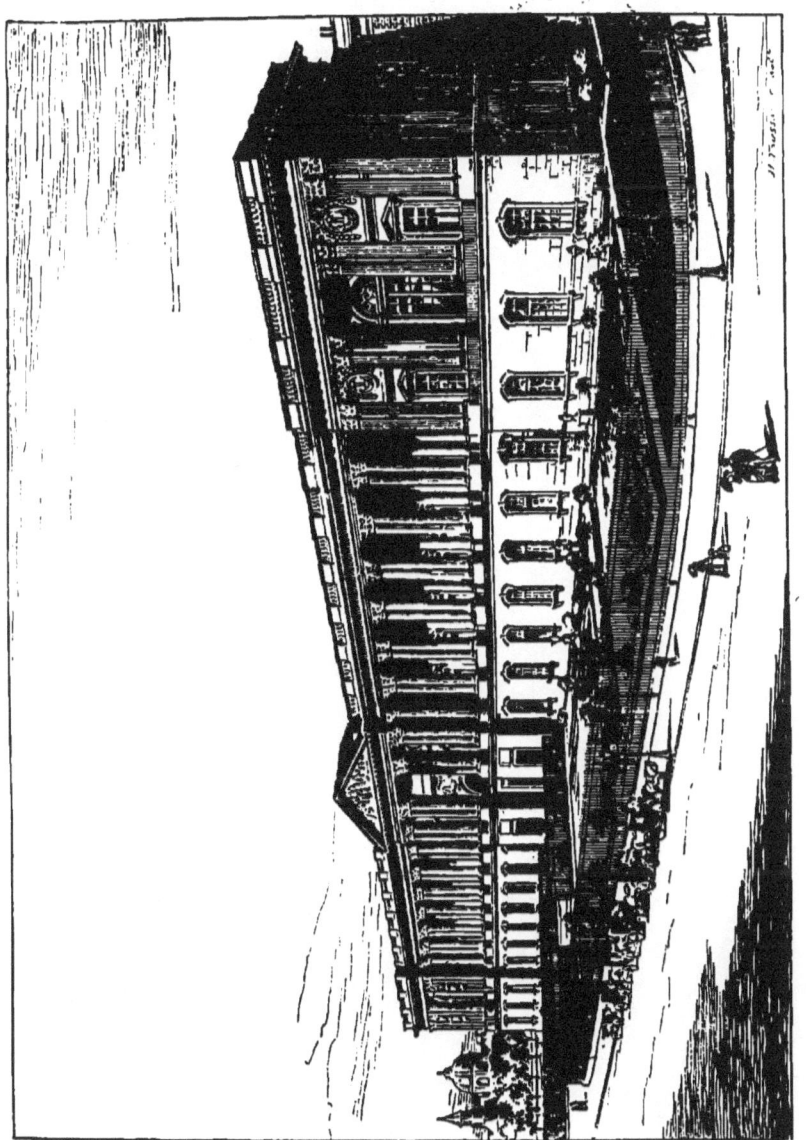

THE COLONNADE. FROM A DRAWING BY H. TOUSSAINT.

each other across the gardens are so near that the spectator cannot retire far enough to see them completely. They can, in fact, only be seen in all their majesty diagonally from the Place du Carrousel. There the effect is stately in the extreme, and very original; there being, I believe, no other palace in the world which offers a perspective of the same kind. Another great merit of the new buildings is that as they enclose a considerable space with their hidden courts and cover a large extent of ground, they *furnish* the space between the Tuileries and Louvre better than some other projects would have furnished it; and this is a merit of some importance, considering the distance between the two palaces. Indeed, Visconti's plan seems to bring the Louvre, by continuing it, as far as the pavilions Turgot and Mollien.

Visconti's buildings have been frequently and severely criticised as "overcharged with ornament." This is an unintentional compliment, for the fact is that his walls are extremely plain, incomparably plainer than the new long gallery of the Louvre, or the new building running east of the Pavillon Marsan. The great effect of richness in Visconti's work is due to the art with which he lavished ornament on certain conspicuous places, especially on his pavilions. A juster criticism is that his work is heavy. No doubt it is massive rather than graceful, but its appearance of enduring strength is not out of place in a public edifice; and though some parts of the old Louvre are more delicate and charming, none are more imposing. The abundance of statues has been blamed, but they are not more numerous than in mediaeval architecture, and they are better detached.

A simpler plan than that adopted by Visconti would have been to dissimulate the want of parallelism between the palaces by making two or three large quadrangles, and losing the radiation in the thickness of the buildings between; but such a plan would have lost the majestic effect of space and distance which it was Visconti's desire to preserve. By his plan the pavilion of the old Louvre could be seen distinctly from the central pavilion of the Tuileries.

The united palaces make so vast a building, that it has been found necessary to give a distinct interest to certain parts. Thus the openings towards the Pont des Saints Pères, called *Les Guichets des Saints Pères*, form an architectural composition in themselves; and that part of Visconti's Louvre which is opposite the Palais Royal is a distinct work, composed for that place and not repeated elsewhere. It is highly ornamented, and contrasts strongly in this respect with the simple work on each side of it.

The sums of money expended on the Louvre and Tuileries defy all calculation. The palaces have not been erected according to any sound principles of economy, but by a system of additions and alterations involving immense sacrifices. As the old castle was pulled down before it was really decayed, so many parts of the Louvre and Tuileries have been replaced prematurely. The river front erected by Le Vau and masked by Perrault is a case in point. Even the long

PERRAULT'S COLONNADE. INTERIOR VIEW.

gallery and the Pavillon de Flore erected by Henri IV. cannot be considered to have lasted very long, as they had to be rebuilt in our own time. The greatest spender on these palaces was Napoleon III. Visconti's plans,[1] when finished by Lefuel, had cost him sums greatly exceeding the first estimate of a million sterling. I believe that the total expenditure on the palaces in our time has reached at least four millions; and if the older work could be accurately estimated in our money it would be equally costly. The total value of the palaces before the destruction of the Tuileries can scarcely have been less than ten millions sterling without their contents; and the value of the site, with its vast area in the best part of Paris, is prodigious.

I have little space to speak of the interior, and it is not a part of my plan to attempt any description of works of art other than architectural. Many rooms in the Louvre are simply plain receptacles for interesting things, but others are interesting in themselves, especially the old wainscoted rooms lined with delicately wrought wood-work from the chambers of the kings. The most sumptuous room is perhaps the Galerie d'Apollon, with its elaborate ceiling, its tapestries in panels, and its collection of precious objects; but the most imposing is the lofty *salon carré*, gravely magnificent, and realizing the grand ideas of Henri IV. As for the long gallery, it is too long to produce its due effect upon the mind, which would be equally potent if it were considerably shorter. It appears to be simply a

[1] Visconti died suddenly in his carriage in 1853.

very magnificent tunnel with pictures on the sides, and nothing near enough to be really visible at the ends. The mere sensation of being in an almost endless tunnel has a distracting effect upon the mind. A room of moderate dimensions, with a few pictures well isolated and well lighted, is much more favorable to the concentration of the faculties in study. The clever comic sketcher Robida has shown us the tramway which, according to him, will be established in that gallery next century. The idea is not unreasonable. A neat little carriage on rails, arranged like an Irish jaunting-car, would be a great convenience for the thousands of tourists who now wearily plod from end to end of that gilded and painted tunnel, with minds distraught and eyes that gaze on vacancy.

AN OLD ROOM IN THE LOUVRE.

VII.

THE HÔTEL DE VILLE.

JUST at this present time (1885) the Parisian Hôtel de Ville seems the most perfectly beautiful of modern edifices, not only on account of the grace and interest of its design, but also because the materials are so irreproachable in their freshness and purity. It would be bold to assert such a thing positively, but it is very likely to be the simple truth that this building, just at present, is the fairest palace ever erected in the world. The reasons why this is likely to be true are the following. To be as perfect as the Hôtel de Ville is now, a building must be erected all together and with a certain rapidity; but great edifices have usually come into being by fragments, so that the parts first erected had time to get old, dingy, and even ruinous before the plan was completed, while the modifications introduced by successive architects have in most cases been fatal to the unity of the work. I need not go farther for examples than to the two great Parisian palaces that we have already studied. Neither the Louvre nor the Tuileries was ever seen as their first architects intended them to be. The palace of the Tuileries, in the whole course of its existence, was never at any time a complete and harmonious work. When it was harmonious (in the time of Catherine de Medicis) it was incomplete, merely a beginning, and when it was complete (in the time of Louis-Philippe) it had long since ceased to be consistent and harmonious. The Louvre is better, but still it is a combination of three or four different architectural schemes, and it is spoiled externally, as a work of art, by being tacked on to a larger edifice, or collection of edifices. Now although the ruder kinds of architecture admit of an unlimited jumble of additions, it is not so with the more refined. The highest kinds of architecture approach, in the strictness of their organization, to the higher animal forms. You cannot give an animal another limb, nor fasten him by suture to another animal, without producing a monstrosity like a five-legged calf or the Siamese twins. So it is in classical architecture of the best kind, and even (though not quite to the same degree) in the best Renaissance architecture. In Gothic, the virtue of unity has been less valued, for the Gothic architects themselves freely added excrescences to their buildings; yet whenever even a Gothic work is in itself exquisitely complete, it cannot be so dealt with except at the cost of that exquisite completeness. Any addition to the Sainte Chapelle would be the destruction of its peculiar beauty.

Now the present Hôtel de Ville (though the design, as I shall show presently, is a growth from an earlier design) is in itself a complete architectural conception carried out at once in all its parts. It is not, like the Tuileries of Philibert Delorme, a beautiful scheme spoiled before it was realized. And the material performance answers in all respects to the idea. The workmanship throughout is of that extreme perfection which is the pride of Parisian craftsmen. The stone is, just now, as fair and immaculate as a selected piece of Parian marble. It is almost as white as snow, and as faultless. It takes the most delicate sculpture as if it were a fine-grained wood, and the quality of its grain is so equal that an artist might sketch upon it as on drawing-paper. The only reproach that can be made against it is that the tone of the whole building is cold; but it is hardly so in sunshine, and there is a beginning of mosaic decoration which promises enrichment of the only kind admissible on so delicate a structure. But not only is the stonework everywhere of the fairest and best, the roofs are perfect to the smallest ornament; and so elegant that although the building is on a great scale it seems more beautiful than vast, and impresses rather by an air of distinction, of aristocracy even, than by any display of power and wealth. It may seem strange to speak of aristocracy in connection with an edifice that is the very centre and council-hall of a mighty and sometimes turbulent democracy; but the word is not misapplied, from an artistic point of view, to a building so completely under the government and discipline of the best architectural authority, having under its command the best and most intelligently obedient labor. Such a building has no natural connection with tumult and disorder. The powers of anarchy did not produce it, could not have produced it. Nor is it

either the product of Philistine wealth. The cost of it
will be about a million and a quarter sterling, yet it
only comes to us as an afterthought that so much good
work is costly. There is sometimes more of the self-
assertion of *bourgeois* money in a citizen's private house
than there is in this great palace. Ornament has been
used sparingly, and what there is of it is chiefly figure-
sculpture. The panels in the front are not carved but
simply divided by mouldings, lozenge-shaped or circular. The *consoles* under the niches between the windows of the central pavilion are very delicately carved,
but the wall behind them is perfectly plain, and the
windows themselves are surrounded by very simple
mouldings. There is a little carving on the two taller
pavilions on each side. Over the arches of the two
beautiful dormer windows, near the clock, there is some
graceful figure-sculpture; and above and about the
clock itself is a fine central composition with colossal
figures and a pediment with the ship of Paris. Yet
even in this, the richest and most central part of the
whole edifice, the ornament is by no means overcharged,
and the figures are relieved by plain spaces of masonry,
as a drawing is by its margin. Among the ornaments
of the roof the most romantic are the men in armor,
with lances, who stand on pedestals along the ridge.
They are gilded, and produce a brilliant effect in strong
sunshine, besides recalling the times when the Hôtel de
Ville was first erected. There are ten of them all
together, — six on the central pavilion, and two on each
of the pavilions to right and left.

FRONT OF THE HÔTEL DE VILLE IN THE TIME OF LOUIS XIII.

The Hôtel de Ville.

It is very commonly supposed that a building has little influence upon the mind when it has no historical associations, but in the case of the present Hôtel de Ville the gain is greater than the loss. It is a virgin building as yet, and may be judged fairly on its merits as a beautiful work of art. It is simply a palace which looks as if it were awaiting the arrival of a prince in a fairy-tale. It seems far too delicate to be in the midst of a populace like that of Paris; and one who loves architecture can scarcely help wishing that it might be transported by magic some night far away in the woods and be safe from bullets and incendiarism. The ways by which a people attains to municipal liberty and parliamentary government are often so rough that the recollection of them gives pleasure only to the enemies of both. If the present building has no splendid memories, if it has received no sovereign within its walls, and been the scene of no extravagant entertainments, it is, at the same time, absolutely free from all revolting and horrible associations. No stormy councils have been begun in its chambers to end in bloodshed; no murder has been perpetrated on its threshold, nor have privileged spectators ever enjoyed from its windows the burning of heretics at the stake, or seen criminals torn limb from limb by four infuriated horses. And not only is the present edifice free from the horrors of history, but it is also free from its vulgarities. The wretched quarrels of yelling demagogues, jumping on tables and crushing pens and inkstands under their heels, have not, as yet, resounded in a building that seems fit only for the presence of gentlemen.

The present building is in its main features a reproduction of that which existed before 1871, but it is not a slavish reproduction; and a comparison between the two shows that the architect took the opportunity for introducing many improvements. What has been done may be explained to a certain extent as follows. Suppose that an artist makes a drawing, well composed,

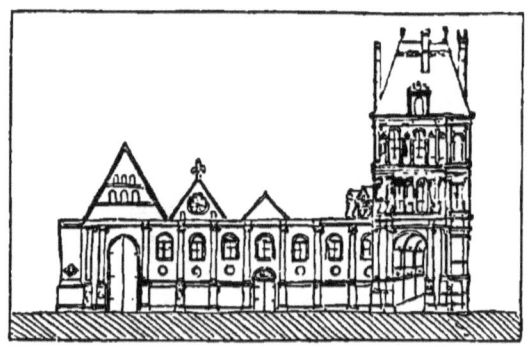

THE HÔTEL DE VILLE IN 1583. FROM A DRAWING BY JACQUES CELLIER.

and in good general proportions, but still leaving room for improvement in other ways; and then suppose that an artist of riper knowledge and more cultivated taste goes over the drawing, pencil in hand, and shows how the ideal which the first artist had in view may be approached more closely. He finds excellent intentions, to which full justice has not always been done. He says, " You might have made more of this idea;

you intended this part of your composition to be elegant, — it may be made more elegant still; these details might be enriched, though without deviating from your intention;" and while he talks in this way he revises the whole work with his pencil; and somehow, without making any very obvious alteration, he gives it greater refinement, and makes it hold better together. I have not space to show in all ways how this has been done in the new Hôtel de Ville, but I may mention one or two instances. The gateway pavilions (those that rise on each side of the central mass) had each of them a sort of encorbelled turret or bartizan, which, with excellent artistic judgment, had been placed to the right in one instance, and to the left in the other, so as to make each pavilion intentionally lopsided and unsymmetrical in itself, yet forming an imperfect part of a perfect whole. The first architect had the idea, which was excellent, but he strangely failed to make the most of it. He diminished the size of the turret in its uppermost story and gave it no roof! It is wonderful that he should have missed such an opportunity. The architect of the new building has been careful not to miss it. He has carried the turrets up to the full height of the pavilions, and then given to each of them a delightfully elegant little roof of its own, carefully finished with an ornamental ridge and finials so as to avoid a pyramidal point, and imitate in little the roofs of the great pavilions. These turrets now occupy the same position that pretty children have in a family, and they give a charm and lightness to the whole edifice that could have been attained by no other means. Again, in the ornamental structure about the clock, and in the bell-turret, the architect has taken the old motives and made more of them. After every allowance has been made for the imperfect draughtsmanship of old engravers, it is evident from their testimony that these important and central parts of the Hôtel de Ville, though the same in general intention as at present, were in old times much less elegant than they are now; and we know from drawings and photographs, if personal recollection were insufficient, that many small improvements upon the edifice as it existed immediately before the Commune have been unobtrusively but effectively introduced into the new design. The corner pavilions are better finished than they were under Louis Napoleon, and so it is all over the building. The intention has been to preserve the traditional forms, but quietly to take every opportunity of improving them. It is a new edition of an old book, not revised by the author, but by a respectful editor more skilful than the author himself.

It is curious that the front of the edifice, which seems to us so happily designed, should be the result of accident. The original plan included only the central mass with the clock and the bell-turret, and the two pavilions which flank it. The design was very pretty and complete in itself; but it was not imposing by its size: and even such as it was the town had the greatest difficulty in carrying it into execution, and it lingered from reign to reign. Francis I. planned the Renaissance edifice; but although he employed a hundred workmen upon it, afterwards reduced to fifty, it was not very forward when he died. It was not finished even at the death of Henri IV. The building was in a very imperfect state for seventy-two years, and remained imperfect afterwards. Nothing proves more clearly the immense inferiority of old to modern Paris in productive power, than the great difficulty experienced by the sovereigns and people of former times in getting forward with their architectural undertakings, which seem in almost every instance, except that of the Sainte Chapelle, to have been far too heavy for their resources. To the modern municipality the erection of such a building as the old Hôtel de Ville would be a small matter. The present one, which has grown from its foundations in the lifetime of a child, is three or four times as vast as that which existed in the imagination of Francis I., and which he could not realize.

The Hôtel de Ville, the Tuileries, and the Palace of the Luxembourg, are all instances of enlarged buildings. If the reader has perused the article on those palaces, he will have observed that they were enlarged in different ways. The Tuileries grew by the addition of masses and pavilions, first on one side then on the other, and all (except the very earliest) out of proportion with the centre, which had to be enlarged afterwards. Then came a general levelling-up and *alignement*, the consequence being a piece of patchwork and mending which never presented the appearance of an artistic composition. The Luxembourg was enlarged in another way. It was already overloaded at one end by four heavy pavilions which stood too near each other, when Louis-Philippe, to get more internal accommodation, made the four into six by adding two others and advancing the front, thereby considerably increasing the defect of heaviness. In the case of the Hôtel de Ville, on the contrary, the enlargement by the addition of masses of building to right and left, set a little back, and pavilions at the corners, coming forward, was done so judiciously, and with such a fine sense of what is suitable and proportionate in a great edifice, that although the present architects had the opportunity of substituting a design conceived all at once, they have been perfectly satisfied with reproducing all the main features of the old building with its appendices. The truth is, that nobody could possibly know, unless he was told, that the wings were additions or appendices at all. It is the happiest instance of successful enlargement that I ever met with. In the interior the increase of dimensions was carried out by the addition of two new courts, one on each side the central quadrangle. All these courts in the new building are exquisitely finished. The two lateral ones have beautiful winding staircases, rich in sculpture, with open balusters and turret-roofs,— an idea which has descended from Gothic times and been adopted by the Renaissance with the addition of elegant ornament. The central court is on a higher level (access to it is had by stairs from the side-courts and the vestibule), and on occasions of great festivity it will probably be converted into a vast hall by the addition of a tent-roof.

The festivities at the Hôtel de Ville have long been celebrated for the combination of magnificence with good taste. The present writer remembers seeing the old building at its best many years ago at a grand ball given by the Municipality to Napoleon III. and Victor Emmanuel. He happened to be in the great court when the sovereigns ascended the stairs, and the combination of beautiful architecture with rich draperies, abundant illumination, and splendid costumes, made a spectacle hardly to be rivalled elsewhere, except in some Italian palaces. The scene in the great gallery was as splendid, but not so entirely outside of the commonplace. The great gallery was converted for a short time into a throne-room; and I happened to be at a little distance from the thrones on which sat the two potentates, — one of them at that time the most dreaded of European majesties, the other only king of Sardinia, a petty sovereign who had won recognition by sending troops to the Crimean war. The guests formed a lane all down the room, and the personages walked slowly along it, greeting those they knew. Since that night what changes! The palace they came from is now the last remnant of a ruin; the municipal palace, then thronged by a crowd of guests, has since been reduced to ashes and replaced by an entirely new structure. The great Emperor, after defeat and humiliation, lies embalmed in a sarcophagus in England, the young hope of his dynasty by his side, and the prince whom he then patronized sleeps royally in the Pantheon at Rome, the first of the kings of Italy. The lives of both have now receded completely into the domain of history, and are as sure to be remembered in future ages as those of any other famous personages who have visited the old Hôtel de Ville. Italy will never forget the rough but good-natured and hearty soldier who so often sacrificed his simple personal tastes to the duties of a more and more exalted station; nor is France ever likely either to forget or forgive the statesman, at one time considered so astute, the ultimate outcome of whose deep-laid schemes was the aggrandizement of her neighbors and the humiliation of herself. There are a hundred other associations with the Hôtel de Ville, which it would be easy to enumerate, but these are among the most recent. If the Republic lasts, it is not very probable that the new building will often be enlivened by the presence of crowned heads; but the municipality will at least be able to hold its sittings without the uncomfortable anticipation of those requests for money which so frequently came from the French sovereigns to the provosts of Paris and the *échevins* of old. The only real inconveniences from which the modern municipality is ever likely to suffer are the excess of its own power and the temptations to its abuse. The Municipal Council has such great resources that it is constantly tempted to place itself in antagonism to the State. The two never work smoothly together for very long, and

THE GREAT BALL-ROOM.

the notice of civic independence has taken such deep root in many minds that they are always ready to see infringements of it in the most ordinary acts of the National Government.

Whatever of evil there may be in our own time, whatever evil deeds may have been done during the Commune, men are certainly less barbarous than they were four or five hundred years ago. Executions are less cruel, prisoners are treated with more humanity. I have passed rapidly over the executions which took place formerly in the Place de Grève, the open space just before the Hôtel de Ville, where they are making the new garden-squares, and where boys amuse themselves with bicycles on the smooth asphaltum; but if the reader wishes to thrill his nerves with horror, he will find nothing more terrible than the deliberate cruelty of those executions in old times; the simple murder by a discharge of musketry under the Commune was tender mercy in comparison. Our warfare, too, barbarous as it still remains, is not quite so horrible as in the middle of the fifteenth century, when hundreds of English prisoners were thrown into the Seine near the Hôtel de Ville, with bound hands and feet, and drowned there in the Seine before the eyes of an unprotesting populace. Let us confess frankly that, notwithstanding all the picturesque interest of past times so delightful to novelists and painters, they are terrible if studied seriously, — terrible if once we realize what they were; and there is no place in the world where we feel this horror of the past more strongly than on the Place de Grève, just before the lovely modern palace which I have been trying to describe. The horror of that dreadful night in May, 1871, when the whole edifice and the houses opposite were in flames, does not really equal the horror of one quiet execution in the feudal times. The destruction of a certain amount of property however valuable, the loss of a certain number of lives in a street battle however passionate and sanguinary the conflict, are less odious than the quiet application of vindictive torture to a single unresisting victim. There are places in Europe where our best charity to the past is to forget it if we can, and this is one of them. Let us look hopefully to the future; and may this, the fairest municipal palace in the whole world, hear no harsher sounds than the discussions of citizens in council, and see no fiercer flame than the light of its own festal illuminations.

VIII.

THE PANTHEON, THE INVALIDES, AND THE MADELEINE.

AS in a former article the two principal Gothic edifices in Paris were studied together, so in the present case the reader is invited to consider three of the principal Renaissance buildings at the same time. The first of these is a church, which has been employed alternately for divine worship and as a Walhalla for illustrious Frenchmen, and which to the present day bears traces of both uses; the second is a church which has become a mausoleum exclusively associated in the popular mind with a great renown entirely unforeseen when the building was erected; the third, again, was begun as a church, continued with the intention of making it a temple for military commemorations, and finally used for ecclesiastical purposes, while still preserving the external appearance of a Greek temple, modified by Roman and Gallic imitation. All these edifices have thus been strangely connected both with religion and with the vanities of human celebrity. All of them, again, have a similar architectural interest as modern experiments with antique architectural forms.

It is one of the commonest of errors, among people who do not trouble themselves to keep chronology in mind, to connect Gothic architecture with Christianity by such an intimate association that they can hardly separate the two. Pointed arches and painted windows appear to them ecclesiastical and even religious, while classical architecture seems much more suitable for lay purposes. Nobody who has this prejudice can regard a Renaissance church with any fairness. The forms of the architecture in Renaissance churches are not *exactly* those with which the early Christians were familiar, but they are incomparably nearer to them than the Gothic forms. As Gothic work looks very old and ruinous (when it has not undergone restoration), we vaguely give it credit for great antiquity, while the real reason for its ancient appearance is because it is an exceedingly frail and unsubstantial kind of architecture, which, after a short time, requires incessant repair. If you divide in three parts the centuries which have passed since the foundation of Christianity, you will place Gothic architecture, a French invention, in the third. It is, in fact, the most modern of all the really original styles, and one which was never associated with the early history of Christianity. There is, consequently, no religious reason for the preference of Gothic architecture for churches, unless it is found that pointed arches are more favorable to religious feelings than round ones, and the various fanciful columns and capitals of the Gothic builders more serious than the limited but well-studied inventions of the Greeks.

The idea of the dome came to France from Italy, and it is unnecessary in this place to trace the architectural

The Pantheon, Invalides, and Madeleine. 141

pedigree of the French Pantheon beyond its ancestor, St. Peter's at Rome, the common inspirer of western imitations. Soufflot, the architect of the Pantheon, was one of those narrow-minded artists who identify themselves completely with a certain phase of art, and who, perhaps, by that concentration of their faculties, express themselves in it as naturally as in their native language. Soufflot committed terrible havoc in Notre Dame, and proved to all future ages that he had neither knowledge nor feeling about Gothic; but when, in 1764, he began the church of St. Geneviève, he had found congenial occupation. The foundation-stone was laid by Louis XV. with a votive intention; but the building was completed in the beginning of the Revolution, and the Constituent Assembly opened it as a "Pantheon," or temple dedicated to all gods, including by extension all heroes or great men. The well-known inscription then placed in large letters upon the frieze over the portico, "AUX GRANDS HOMMES LA PATRIE RECONNAISSANTE," is a clear explanation of the sense attached to the Greek name of the building; and a very fine inscription it is, saying all that is needed in six perfectly cadenced words, full of noble purpose and patriotic feeling. Louis XVIII. handed over the edifice to the clergy in fulfilment of the original intention of Louis XV., and it remained in their hands, with the inscription effaced, until the revolution of July. Under Louis-Philippe and the Second Republic it was a Pantheon again, with the inscription restored; but on the establishment of Louis Napoleon's personal power, when he was buying the support of the clergy, the Pantheon was given to them a second time, and they were allowed to keep it until the funeral of Victor Hugo. They re-established altars in the interior, and a cross upon the

THE PANTHEON.

dome, but they did not efface the inscription. In the early revolutionary stage of the Third Republic, there being some apprehension that the Pantheon might be secularized again, a plan was matured for its decoration

The Pantheon, Invalides, and Madeleine. 143

with religious paintings as a sort of final *prise de possession* in favor of the Church; but this has availed nothing, and now (1885) it seems possible that the paintings may be removed. This could be done, I believe, without destroying them.[1]

Much opprobrium has been cast upon the Republican Government for its conduct in this matter, but it may be remembered that a monarch, Louis-Philippe, did exactly the same thing; and if consecration is of eternal effect, then the English noblemen who have turned old abbeys into luxurious country-houses must be equally culpable. The Pantheon has never been a parish church, and the persons whose desires or interests have been most interfered with are a dead king and a saint who died in the early twilight of French history.

The Pantheon has stood the test of a hundred years of criticism, without which no building can be considered sure of permanent fame. Its merits are not of a kind to excite enthusiasm, but they gain upon us with time, and satisfy the reason if they do not awaken the imagination. We can never feel with regard to a severe classical building like the Pantheon the glow of romantic pleasure which fills sense and spirit in Notre Dame or the Sainte Chapelle. If there is emotion here it is of a different kind. The building has a stately and severe dignity; it is at once grave and elegant, but it is neither amusing as Gothic architecture often is by its

[1] As the removal of the paintings is uncertain, the account of some of them which appeared in the first edition of this book is retained.

variety, nor astonishing as Gothic buildings are by the boldness with which they seem to contravene the ordinary conditions of matter. The edifice consists of a very plain building in the form of a cross, with a pediment on pillars at one end and a dome rising in the middle. There are no visible windows, a renunciation that adds immensely to the severity and gravity of the composition, while it enhances the value of the columns and pediment, and gives (by contrast) great additional lightness and beauty to the admirable colonnade beneath the dome. There does not exist, in modern architecture, a more striking example of the value of a blank wall. The vast plain spaces are overwhelming when seen near, and positively required the little decoration which, in the shape of festooned garlands, relieves their upper portion. At a little distance the building is seen to be, for the dome, what a pedestal is for a statue; and the projection of the transepts on each side of the portico, when the edifice is seen in front, acts as margin to an engraving. Had their plain surfaces been enriched and varied with windows, the front view would have lost half its meaning; the richness of the Corinthian capitals and sculptured tympanum, and the importance of the simple inscription, draw the eye to themselves at once.

The situation of the Pantheon is the finest in Paris for an edifice of that kind. Only one other is comparable to it, Montmartre, on which is now slowly rising a church of another order, dedicated to the *Sacré Cœur*. The dome of the Pantheon is one of the great land-

The Pantheon, Invalides, and Madeleine. 145

marks of Paris; it is visible from every height and from a thousand places of no particular elevation. It does not simply belong to its own quarter, but to the whole city.

The interior is interesting in different ways, both as an experiment in architecture and as an experiment in the employment of mural painting on an important scale. The first point likely to interest an architectural student is the manner in which the architect has combined his vaults and his pillars. Soufflot's tendency (unlike that of the architects of St. Peter's in Rome and St. Paul's in London) was towards an excessive lightness. His project was to erect his dome on elegant pillars; but these were found insufficient, and another architect (Rondelet) replaced them by massive piers of masonry. Elsewhere there are Corinthian columns carrying a frieze and cornice, and above the cornice a groined (intersected) vault, of course with round arches, and having exceedingly slender terminations, as this system of vaulting cuts away nearly everything and leaves a minimum of substance at the corners to bear the weight. You may see such vaults frequently in the works of the early Italian painters, but they always support them by very slender and elegant columns; whereas in Soufflot's work they rest on a Corinthian order, with its entablature, which gives the idea of a contradiction, for either the vaulting is too light or the entablature is needlessly heavy. The Italian painters were consistent on the side of lightness, Wren on the side of heaviness; but it seems as if Soufflot had rather confounded the two, so far as the satisfaction of the eye is concerned.

There is a remarkable peculiarity about the level of the floor; the aisles and transepts are higher than

THE PANTHEON FROM THE GARDENS OF THE LUXEMBOURG.

the nave, into which you have to descend by five steps. The general aspect of the interior is agreeable, from the pleasant natural color of the stone and its thoroughly careful finish everywhere; but the large spaces of wall, though divided by half-columns, were felt to be too bare, and there have been various projects for their decoration. That which is now being carried into execution includes the painting of many mural pictures at a height which we should describe as the line in an exhibition, and also of decorative friezes at a greater height above the eye. I have mentioned the columns which, half buried in the wall, divide what, without them, would be its too extensive spaces. The existence of these columns cuts the wall into a series of upright panels not always convenient for the purposes of an artist, so it has been decided that the larger compositions should include three of these spaces, and that the picture should in these cases appear as if it were seen behind the columns, which themselves are left without any kind of painting or decoration. The plan was the best that could have been adopted under the circumstances, as the artists would have felt cramped by being confined to narrow upright panels; but it required very careful management to preserve Soufflot's architectural effect.

Mural painting ought never to make us feel as if the wall were taken away, because that is an injury to the architecture. The painting should be so far removed from realism that we feel the wall to be a wall still, upon which certain events have been commemorated. Among French mural painters, not one has understood this so well as Puvis de Chavannes, and it would have been wise to entrust to him the entire decoration of the Pantheon, both for the sake of the architecture and for the unity of the work; but, unfortunately (so far as these considerations are concerned), other men have also been called in, men of great ability, no doubt, yet who were not disposed to make the necessary sacrifices. Puvis de Chavannes is essentially a mural painter. He has accepted the conventionalisms of that kind of art, and his mind is so exceptionally constituted that such restraints are evidently agreeable to him and favorable to his inventive powers. His large work in the Pantheon represents the finding of Saint Geneviève when a child by Saint Germain and Saint Loup, at Nanterre, when they were journeying towards England. The bishop sees that the child has a religious aspect, "has the Divine seal upon her," and predicts for her a memorable future. This takes place in a vast landscape, with undulating ground and fine trees in the middle distance against a line of blue hills, and a blue sky with white, long clouds. In the foreground is a rustic scene, including the milking of a cow under a shed; and in the middle distance we have a view of Nanterre, or at least of a mediaeval city. The figures are all very simply painted in dead color, kept generally pale and hardly going beyond tints, which are often false so far as nature is concerned, but never discordant. Such painting is very reticent, very consistent; and, though it is not true, it contains a great amount of truth, and implies far more knowledge than it directly expresses. The landscape background, for example, is simple, but it is not ignorant; it shows quite plainly that the painter is a man of our own century, perfectly conversant with our knowledge, yet decided not to go beyond a certain fixed point in the direction of actual imitation. The figures are exceedingly dignified; but when the painter gets away from the muscular type, and has to deal with weaker men or with children, he is not so satisfying. A smaller picture represents the child Saint Geneviève praying in a field, while the rustics watch and admire her. The sentiment here is very pure and simple, like that of an idyllic poem. In the upper part of the composition a ploughboy, behind trees, watches the saint while his oxen rest; in the lower part, a peasant man and a woman watch her also.

Now, although these paintings tell their story perfectly, not a single person or other object in them is so far realized as to make us forget the wall-surface. A story has been told upon the wall just as an inscription might have been written upon it, but nothing has been done to take the wall away. Even the pale tinting is so contrived as not to contrast too violently with the natural stone around it. Let the visitor who has just seen these paintings, and, perhaps, been a little put out by their conventionalism, glance up from them to the pendentives under the dome painted by Carvallo from drawings by Gérard. Those works are strong in darks, and in far more powerful relief than the situation warrants. They are also surrounded by heavily gilt carvings, which make the surrounding stone look poor; in short, from the architectural point of view, they are a series of vulgar blunders. I would not use language of this kind with reference to so serious, so noble an artist as Jean Paul Laurens, but I cannot help regretting that his magnificent composition of the death of Saint Geneviève was not in some public gallery rather than in the Pantheon. The realization is far too powerful for mural painting. We do not see a record on a wall, but the wall is demolished, and through the opening we witness the scene itself, the infinitely pathetic closing scene at the end of a saintly life, when, even in the last moments of extremest weakness, a venerable woman still throws into the expression of her countenance the benedictions that she cannot utter. One consequence of the external force with which all the figures and objects are realized in full modelling and color is that the two columns which cross the work vertically are felt to be in the way; in other words, the architecture of the Pantheon is in the way, and so far from helping the architect, the painter has done him an injury, for what are smoothly chiselled stones, what are fluted columns and pretty Corinthian capitals, to the awful approach of Death?

On the other mural paintings in the Pantheon we have no need to dwell. So far as I know them yet[1] they belong to the class of historical genre common in the French salons, and have neither the power of Laurens nor the careful adaptation of Puvis de Chavannes. Cabanel's pictures represent three scenes in the history of Saint Louis,—one his childhood, when

[1] Some paintings on the south side have been uncovered lately, and these I have not seen.

he is being taught by his mother; a second, his civil justice; and a third, his military life as a Crusader. The first subject is the best suited to Cabanel's talent, and is a pretty domestic scene. The subject selected by M. Maillot for his paintings in the south transept is a mediaeval procession with the relics of Saint Geneviève, and these paintings are a good example of a danger different from the powerful realization of Laurens. In the present instance the evil is a crudity of brilliant color, like mediaeval illumination, which always seems out of place on a wall unless it is carried out consistently by polychromatic decoration throughout the building.

It is sometimes said by journalists that these paintings are frescos (wall-paintings are generally taken for frescos). The fact is that they are oil-paintings on *toile marouflée*, that is, on canvas fastened to the wall by a thick coat of white-lead. This is now the accepted method for mural painting in France. It is convenient for the artist, as it allows him to paint in his own studio in a material he is accustomed to use; and it is believed to be as permanent as any other.

The dome of the Pantheon attracts the eye simply by its own architectural beauty; but that of the Invalides, by Mansard, is lustrous with abundant gilding, and on a sunny day shines over Paris with the most brilliant effect. It is splendid against one of those cerulean skies that are still possible in the capital of France. Certainly nothing does so much for the splendor of a great city as very conspicuous gilding. There are drives in Paris, as, for instance, from the Trocadero to the Place de la Concorde, during which the dome of the Invalides accompanies you like a harvest-moon. On a nearer approach it is the architecture that claims attention. The dome itself is fine, but in many re-

THE INVALIDES.

spects the building as a whole is greatly inferior to the Pantheon. Soufflot made the body of his church an ample base for his dome in every direction; but at the Invalides one receives the impression of a man with a prodigious head on a small body and very narrow shoulders. The columns of the dome are in couples,

The Pantheon, Invalides, and Madeleine. 153

with projecting masses doing the work of buttresses. This gives more light and shade than the simple colonnade of the Pantheon, but not such beautiful perspective, as the projections interfere with it. The composition of the front makes us feel strongly the special merits of the Pantheon. Instead of the majestic columns of Soufflot's work, his rich pediment, and the massive plain walls on each side as margin, we have in the Invalides a poor little pediment reduced to still more complete insignificance by the obtrusive windows, etc., on each side of it. Again, the front of the Invalides offers an example of that vice in Renaissance architecture which Soufflot avoided, — the superposition of different orders. It is divided into two stories, Roman Doric below and Corinthian above, a variety that the Renaissance architects enjoyed, though it does not seem more desirable than two languages in one poem.

This criticism does not affect either the beauty of Mansard's dome as a fine object seen from a distance, or the importance of the interior, one of the most impressive in all Paris, especially since it has become the mausoleum of Napoleon I.

A lofty dome, supported by massive piers perforated with narrow arched passages and faced with Corinthian columns and pilasters, a marble floor of extraordinary richness and beauty everywhere, all round the base of the dome a stair of six marble steps descending to the circular space under it, and in the midst of this space a great opening or well, with a diameter of more than seventy feet, and a marble parapet, breast-high, for the safety of the visitors who look down into it, — such is the first impression of the interior.

Not only do the people invariably look down, but they generally gaze for a long time, as if they expected something to occur; yet a more unchanging spectacle could not be imagined. In the middle there is a great sarcophagus of polished red Russian granite, and twelve colossal statues stand under the parapet, all turning their grave, impassible faces towards the centre. They are twelve Victories whose names have resounded through the world, and in the spaces between them are sheaves of standards taken in battle, and in the red sarcophagus lies the body of Napoleon.

The idea of this arrangement is due to the architect Visconti, who had to solve the problem how to arrange a tomb of such overwhelming importance without hiding the architecture of so noble an interior as this. His solution was admirably successful. The arrangement does not interfere in the slightest degree with the architecture of the edifice, which would have been half hidden by a colossal tomb on its own floor; while we have only to look over the parapet to be impressed with the grandeur and the poetic suitableness of the plan. With our customs of burial we are all in the habit of looking down into a grave before it is filled up, and the impressiveness of Napoleon's tomb is greatly enhanced by our downward gaze. We feel that, notwithstanding all this magnificence, we are still looking down into a grave, — a large grave with a sarcophagus in it instead of a coffin, but a grave nevertheless. The serious grandeur, the

The Pantheon, Invalides, and Madeleine. 155

stately order, of this arrangement seems to close appropriately the most extraordinary career in history; and yet it is impossible to look upon that sarcophagus without the most discouraging reflections. The most splen-

THE MADELEINE.

did tomb in Europe is the tomb of the most selfish, the most culpably ambitious, the most cynically unscrupulous of men; and the sorrowful reflection is that if he had been honorable, unselfish, unwilling to injure others, he would have died in comparative or in total obscurity, and these prodigious posthumous honors would never have been bestowed upon his memory.[1]

The church of the Magdalen (Madeleine) is curiously connected with the history of Napoleon I., who had the incompleted edifice continued with the strange intention of dedicating it as a temple to the memory of La Grande Armée. Every year, on the anniversaries of the battles of Austerlitz and Jena, the temple was to have been illuminated and a discourse delivered concerning the military virtues, with an eulogy of those who perished in the two battles. This intention was never carried out, and the building, which had been begun in 1764 as a church, was finished as a church under the reign of Louis-Philippe. Nothing could apparently be more decided in architectural intentions than the Madeleine as we see it now. It seems to be plainly a temple, and never to have been intended for anything else. In reality, however, it was begun under Louis XV. as a church, resembling what is now the Pantheon, and the change of plan was carried into effect many years after the works had been actually commenced. It is not by any means a subject of regret that this temple should have been erected in Paris, as it gives many students of architecture who have not visited the south of Europe an excellent opportunity for *feeling* what an antique temple was like, to a degree that is not possible with no

[1] Some fresh example of his baseness is constantly cropping up. During the last visit I paid to the Invalides, in May this year (1883), I could not help thinking all the time about that letter to which Napoleon forged the signature of Davoust, and for publication too, as narrated not very long since in the "Revue des Deux Mondes."

The Pantheon, Invalides, and Madeleine. 157

more powerful teachers than photographs or small models. Viollet-le-Duc said that it was barbarous to build the copy of a Greek temple in Paris or London, or among the mists of Edinburgh, condemning alike the Madeleine and the fragmentary Scottish copy of the Parthenon; but surely a student of architecture, born in the north, would visit both the Scottish Parthenon and the Parisian temple with great interest, simply because they show him columns on their own scale, real columns in the open air. We are so accustomed to Gothic and Renaissance churches that a temple is an acceptable variety, were it only to demonstrate, by actual comparison, the immense superiority of more modern forms for purposes of Christian worship. We ought to bear in mind, however, that although the Madeleine resembles a Corinthian temple externally, it has not the surroundings of such a temple and is not associated with its uses. For Christian architecture, on the other hand, such a system of building involves a great waste of money and space in the colonnades and the passages between them and the walled building or *cella*. The space in the Madeleine, already so restricted, is limited still farther by internal projections intended to divide the length into compartments and to give a reason for six lateral chapels, so that every one who enters it for the first time is surprised by the smallness of the interior. I need hardly observe that there is not the slightest attempt to preserve the internal arrangements of a Greek temple, even if they were precisely known, on which architects are not agreed. The side chapels have arches over them, the roof is vaulted with round arches across the building, springing from Corinthian columns, and in each section is a dome-ceiling with a circular light (as in the Pantheon at Rome), these lights being the only windows in the edifice. The high altar is in a round apse *en cul de four*, with marble panels and a hemicycle of columns behind the altar. There is great profusion of marbles of various kinds, of gilding, and of mural painting, that I have not space to describe in detail. Enough has been said to show that the work, as a whole, is a combination of Greek, Roman, and French ideas. The general idea of the exterior is Greek, but if you examine details you see the influence of Rome, and you find it still more strongly marked inside, by the arches of the roof. The French spirit is shown in the decoration chiefly, which is so truly Parisian that the Madeleine is instinctively preferred by fashionable people. A fashionable marriage there is one of the most thoroughly consistent spectacles to be seen in modern Paris. Here is nothing to remind us of the austerity of past ages, but the gilded youth of to-day may walk along soft carpets, amid an odor of incense and flowers and the sounds of mellifluous music. The pretty ceremony over, they pass out down the carpeted steps, and an admiring crowd watches them into their carriages. And nobody thinks about the dead at Austerlitz and Jena!

IX.

ST. EUSTACHE, ST. ETIENNE DU MONT, AND ST. SULPICE.

NEXT to Notre Dame, St. Eustache is the largest church in Paris, and the difference between them is much more marked in length than in height and breadth. The length of Notre Dame is nearly 127 mètres, that of St. Eustache only 88½; but while the breadth of Notre Dame is 48 mètres, that of St. Eustache is nearly 43; and the difference of height between the two edifices, internally, is scarcely more than one English foot in favor of the Cathedral. Besides their similarity in height and width, the two churches have an important feature in common, — their double aisles. In short, it seems as if the builders of St. Eustache had in their minds some distinct idea of rivalry with Notre Dame, at least to a certain degree.

Before the present church of St. Eustache, there existed a Gothic edifice that was not half so long, nor half so broad either, so that it would not occupy a quarter of the area; and if its height was proportionately small (which is probable, as the present building is very lofty), the cubic dimensions of the old church[1]

[1] There had been another church still earlier, and perhaps a still more remotely ancestral edifice than that; but of these we know nothing.

would be less than one eighth of those of its successor. It is evident, therefore, that so far as the importance of the edifice is concerned we have nothing to regret; and it is not probable that the Gothic church exceeded the present building either in elegance of design or perfection of workmanship, while it may be accepted as certain that it could not have been so interesting to the student of architecture because the St. Eustache that we know is a valuable experiment on a scale sufficiently imposing for it to be really decisive.

The interest of St. Eustache consists in this, that the designer, whoever he may have been, attempted to combine the general impressiveness of a Gothic edifice with the spirit of the Renaissance in every detail. He must have admired Gothic architecture in a certain fashion, and he must have appreciated its influence on the mind, yet at the same time he did not admire it enough to follow it slavishly in anything. Nobody knows who the first architect was. It has been said that his name was David; and there was a Charles David buried in the church, whose epitaph says that he was architect and conductor of the building of that church; but he must have been a successor to the first architect, as the first stone of the present building was laid by the Provost of Paris in the year 1532 (August 19th), while Charles David was born in 1552. It is much to be regretted that the original architect's name should have lapsed into complete oblivion, as he was an original thinker in architecture and a man of poetic imagination.

THE CHURCH OF ST. EUSTACHE.

St. Eustache, St. Etienne, and St. Sulpice. 161

St. Eustache is closely connected, chronologically, with the Hôtel de Ville, as that edifice was begun just a year after the church. It has been supposed that the architect of St. Eustache must have been the architect of the Hôtel de Ville, or else one of his pupils; but this is a mere supposition, without any evidence to support it. We may observe that although both edifices are works of remarkable merit, their merit is not the same. The Hôtel de Ville is simply a Renaissance palace, very beautiful, but not attempting to solve any such problem as the reconciliation of two opposite styles; while the Church of St. Eustache is from beginning to end a sustained and conscious effort to unite the imposing effect of Gothic with the delicacy and comparative sobriety of Renaissance architecture. The result is a hybrid in which every visitor who knows enough about architecture to be able to disentangle two opposite elements will find ample and pleasurable occupation.

The ground-plan of St. Eustache approaches more nearly to that of Notre Dame than would be believed from the outward appearance of the two edifices. At St. Eustache the long limb of the cross is much shorter in proportion; but you have the same four lines of columns, or piers, the same round apse and *pourtour*, and the same series of small chapels outside the double aisles. In both edifices the transepts only reach to the external walls of the chapels.

Other features that the two buildings have in common are flying buttresses, rose-windows in the transepts, and spires at the intersection of the roof. That of Notre Dame has been restored, as we have seen, but the spire of St. Eustache was long since shortened to make a platform for a semaphore telegraph, and has never been re-established.

The comparison fails most decidedly at the west end. Everybody knows that Notre Dame has twin towers and a great west front; but, unfortunately, of the twin towers that St. Eustache was to have had only one has been built, and that is small and not noteworthy. Nor is it really one of the towers intended by the original architect. It is an invention of the eighteenth century, when it was thought necessary to erect a new *portail*, which included a complete new front. The unknown original architect had built a west front completely in harmony with the rest of the edifice; but as for the towers, he had only carried one of them partly towards the height of the first detached story, while the other, though prepared for, was not carried high enough to detach it from the body of the church. Still, though incomplete, the original front was beautiful, being as elegant in its severer parts as the rest of the exterior; while, in obedience to Gothic precedent, it was enriched with statues on the buttresses and in the doorways, and with other decorative sculpture, which, if we may judge by what remains elsewhere, must have been of the most delicate and refined quality. That was in the time of the elegant Renaissance, when fancy and invention were not yet excluded from architecture. Then came the terrible mechanical period in the eighteenth century, when both architects and the public per-

CHURCH OF ST. ETIENNE DU MONT. FROM SKETCH BY A. BRUNET DEBAINES.

St. Eustache, St. Etienne, and St. Sulpice.

suaded themselves that graceful fancy was too light an element to be admitted in serious art; and it happened unfortunately that the west front of St. Eustache was rebuilt during this period, without the slightest consideration for the desire of the original architect that the church should be a combination of Gothic with Renaissance forms.

The new *portail* was a very severe and very dull arrangement of Roman Doric on the ground story, with Roman Ionic and a plain pediment above. The one tower that was built is in a sort of Italian Corinthian. In order that the pediment might not appear too absurdly out of place, the lofty old gable which would have appeared above it was cut off like the side of a pyramid with an Italian balustrade at its base. The general result is a huge *applique* that no more belongs to St. Eustache than it would belong to the Sainte Chapelle. It is much to be regretted that a complete restoration of this part of the church was not undertaken during the reign of Napoleon III., when it might have been quietly carried into effect. At the same time towers might have been built in the spirit of the original work. It is now too late to dream of any such expenditure on the part of the Government; and the priests have enough on their hands with the huge monumental church of the Sacred Heart on Montmartre, which absorbs all the money that can be collected.

It is interesting to observe in what way the classical tastes of a Renaissance architect modified Gothic forms. Greek architecture, though elegant, was stiff and angular; Roman architecture, though less visibly angular because it had the round arch, was still simple and severe; but Gothic architecture became pliant like the branches of trees and lively like tongues of flame. In St. Eustache the Gothic forms are stiffened by classical feeling. The tracery of the windows is simplified and monotonously repeated in corresponding parts of the church. This simplification is especially visible in the rose-windows, so poor and angular in comparison with true Gothic. Again, in the spaces over the doors, instead of the richly inventive sculpture of the Gothic tympanum, with its elaborate story of the Fall of Man or the Last Judgment, the Renaissance architect has introduced hexagonal tracery almost like the cells of a honeycomb. Even in the large pilasters with Corinthian capitals the half-column becomes an elongated panel with a triangle at the top, and another triangle at the bottom, pointing towards each other. For the intricately curved iron-work on Gothic doors we have plain oblong panels giving sixteen right angles to each door. In a frieze running above the lowest windows triglyphs are introduced, and all the rest of the ornamentation is so angular that they do not seem out of place. With its exceedingly perfect finish, and its abundance of plain little details, the outside of St. Eustache reminds one of nothing so much as a masterpiece of serious cabinet-making. And the wonder is, that although the style is a jumble of reminiscences from Greece, Italy, and mediaeval France, not one of them in a condition approaching to purity, the whole is per-

INTERIOR OF ST. ETIENNE DU MONT.

fectly harmonious. The reason is that every borrowed idea has been so modified as to combine with every other.

The interior has one transcendent merit, and several obvious defects. The merit is an overpowering sublimity due to the expression of height which again is in great part the result of the narrow space between the columns, or piers, and the elevation of the point at which the arches spring. It is like being at the bottom of a deep and narrow ravine and seeing it spanned by a little stone bridge far up above our heads. The impression of loftiness is also greatly aided by the unusual height of the aisles.

Unfortunately, the narrowness of the space between the piers, and the comparative massiveness of the piers themselves, have the bad effect, sometimes met with in Gothic churches, of impeding the view diagonally. So long as you are in the large open space of the nave it is well, because that space is open enough to prevent any sense of confinement; but though the aisles are very lofty they convey the feeling of narrow passages, because the successive piers of masonry are joined together in perspective exactly as if they were walls, and you only get a glimpse through the opening which is nearest you. Some readers may remember the remarkable difference in this respect between the Cathedral at Rouen and the well-known Church of St. Ouen in the same city. The Church of St. Ouen is much more open, which gives more spacious perspectives, and may be one of the reasons why it is so generally preferred to the

Cathedral, in spite of some architectural authority on the other side.

There is one notable advantage in the mixed style of St. Eustache. It is near enough to classic architecture to admit without incongruity both learned figure-sculpture and learned modern painting, so that there is no necessity for archaic forms in either. It is probably for this reason that St. Eustache seems more happily and suitably decorated than most other churches.

On the whole, we must come to the conclusion that the interesting experiment of combining Gothic effects with classical details and finish could not have been made more intelligently than here. It is not at all an unreasoned decadence of Gothic; it is a combination at once logical and imaginative. The unknown architect was an artist, and a great artist; he could rise to the sublime, and enjoy the exercise of a delicate and discriminating taste. Yet in spite of his rare powers of combination he founded nothing. The style of St. Eustache might have become the modern style, but it did not. In the eighteenth century men fell into that heavy style of pseudo-classical architecture founded on debased Italian precedent, which mistook dulness for dignity, and of which we have a striking example in the west front of St. Eustache itself. In the nineteenth, ecclesiastical architecture in Paris has gone in two directions, — either towards a revival of past styles, as in the meagre Gothic Church of St. Clotilde, the Gothic Church of St. Bernard (Rue d'Alger) and others, the Romanesque St. Ambroise (Boulevard Voltaire), and St. Pierre de Montrouge; or else towards the invention of a thoroughly modern style, as in St. Augustin, the Trinity, and St. François Xavier. It is useless to indulge in unavailing regret, and it may have been necessary to the full understanding of Gothic by the architects of our time that many of them should pass through that wretched state of probation known by its fruits in miserable pseudo-Gothic; yet it seems as if, in St. Eustache, they had a compromise between modern finish and Gothic invention which might have suited modern capabilities, and at the same time have harmonized with the development of other arts.

The Church of St. Etienne du Mont (near the Pantheon) is not, like St. Eustache, an example of the complete fusion of the Gothic and Renaissance ideas; it is an example of Gothic in its decadence, strongly influenced by Renaissance, and finally lost in the new style from which every trace of Gothic is eliminated. There is, consequently, in St. Etienne nothing of that strong and peculiar artistic interest that belongs to the remarkable edifice we have just been describing. St. Eustache stands alone, but there are many churches in which a debased Gothic is clung to with hesitation, and at length abandoned, in some important part, for the style that had come into fashion. Still, very few of these churches can be compared to St. Etienne for a certain romantic charm. Only the most severe and intolerant purists in Gothic would quarrel with a decadence like this, in which, if a great art is dying, it dies like the last cadences of music, leaving only a regret for the end of inspiring or sweet emotions. You may build a church entirely according to rule, you may copy in all its details the best art of the best time, yet not succeed in awakening any feeling beyond a cold approval of your accuracy. In St. Etienne there are many deviations from precedent, many things that are theoretically difficult to defend; but the building is a poem, the architect was an artist who had feeling and imagination, and this small interior impresses the mind more powerfully than many that are far vaster and incomparably more costly.

We have seen that in St. Eustache the view is diagonally blocked by the nearness and thickness of the piers. In St. Etienne this fault is happily avoided. The architecture is everywhere open and penetrable, and the intersections are delightful, especially because you are always sure to have painted windows in the background. The clerestory is proportionately low, being only the height of the arch in the groined vault itself; and consequently the pillars would have appeared too high had they not been united, at nearly half their height, by a gallery on arches, which is one of the original features of the church. This gallery, which (though otherwise placed) answers to the triforium in pure Gothic edifices, is exceedingly light, with open balustrades, and it has afforded an excuse for the elegant staircases that wind round the columns on each side the beautiful rood-screen, and belong to it, not only by their design, but also as parts of the same beautiful and elaborate composition.

WEST FRONT OF ST. ETIENNE DU MONT.

St. Eustache, St. Etienne, and St. Sulpice.

The great charm of St. Etienne is the beauty and variety of the accidental views in it. There is in every church the great view down the nave, and if that is not successful we say the building is a failure; but besides this supremely important aspect of the building, there is, or there is not, the quality of revealing unexpected beauties. Some churches are very remarkable for the possession of this quality, — they even possess it to a degree that the architect himself may possibly not have foreseen; others are absolutely destitute of it. There is no trace of it in the Madeleine. When you have been in the Madeleine a quarter of an hour you have nothing more to discover as to the possibilities of its architecture, and for any new interest you must turn to the decorative details added by the sculptor and the painter. On the other hand, there are many little-known churches — such as that at Dreux, for example — of which the charm consists in lovely combinations, that seem entirely accidental, and which a painter would immediately select in preference to the long, formal view down the nave. The best places for finding these are near the intersections of the nave and transepts, and in the *pourtour* round the apse, when happily there is one. In St. Etienne du Mont all the necessary conditions for producing happy accidental combinations exist in the utmost perfection. The view is never blocked up, and there is always a rich mystery of painted glass at the end of it, relieving the cool color of the stone. The prettiest of these minor views are those from the aisles looking across the transepts and towards the apse, because there you get the extremely elegant work of the rood-screen, which is continued across the aisles, leaving a passage through beautiful doorways under the prettiest little pediments imaginable, surrounded with fanciful and delicate sculpture in the charming taste of the refined Renaissance.

The west front of St. Etienne is very well known from photographs. It is a curious composition, not defensible, logically, yet picturesque and elegant in the total result. First you have a pediment supported on four imbedded columns of a debased Corinthian, with an arch above the tympanum over a square-headed door. Above the apex of the pediment oddly comes a rose-window, much nearer to pure Gothic than those in St. Eustache, and over the rose-window a *fronton* in the segment of a circle like those which alternate with pediments on the river-front of the Louvre. To crown all, we have a highly pitched gable, essentially Gothic in principle, but with Renaissance ornament. The tower is narrow and elegant, and the composition of the front is happily aided by a little turret with pepper-box roof low down to the left. To a taste educated in the severe tradition either of Greek or of pure Gothic such a combination as this must seem indefensible, yet it is at the same time elegant and picturesque. It may be proved, by reasoning, to be incongruous; and yet there is so much good management in the proportioning of the parts and the finish of the details that it is impossible to turn away from such a work without a tormenting desire to look at it again.

THE CHURCH OF ST. SULPICE.

The Church of St. Sulpice is very imposing from its dimensions and the sober massiveness of its construction, but it has none of the charm which belongs to the two edifices we have just been studying. The front is composed of two stories that include the lower parts of the towers, and between the towers an open portico with a loggia above. The architect employed two orders in the front, — Italian Doric in the lower story and Italian Ionic in the loggia. Corinthian is freely employed in the northern tower, and a sort of Corinthian also in the other, which has never been externally finished, though it has attained its full height. A common criticism of this front is that it does not answer in any special manner to the interior of the church, of which it explains nothing. It is, in fact, only a gigantic screen giving the church a sort of adventitious importance. Architecture of this kind may excite admiration by majesty and grandeur, but, unlike the work of the elegant Renaissance, it can never charm or delight. It is the architecture of pride and power; it is not the architecture of inventive affection.

The rest of St. Sulpice externally is heavy, substantial, and dull. It is, I believe, a most respectable piece of building and likely to be very durable, but it seems destitute of fancy and imagination. The interior has round arches springing from massive piers against each of which is a Corinthian pilaster, and the roof is simply vaulted with a large arch springing from the walls pierced with lower vaults for the clerestory windows. The effect is serious without any of the lightness and grace that characterize the Pantheon. Much of the effect of St. Sulpice is due to its great size. The measures given by different authorities are not precisely alike; but it appears from them that the Church of St. Sulpice is longer and broader than Notre Dame, and very nearly as lofty in the interior. The towers of St. Sulpice are higher by two mètres than those of the metropolitan Cathedral, which they resemble in this, that they were to have had spires, or some kind of superstructure that was never added for fear of insecurity.

The greatest artistic attraction in St. Sulpice is the chapel of the Holy Angels, with three large mural paintings by Eugène Delacroix. He painted these on the wall itself, which he primed in white lead with his own hand. They were finished in June, 1861, and Delacroix admitted people to see the chapel by circular before it was open to the general public. He was anxious about the effect on the art world, and rather disappointed, as M. Charles Blanc and others were decidedly cool, and the press was much divided in opinion. Since then Delacroix is better understood, and we are not so much disconcerted by his violent action and strong coloring. The subjects on the walls are Heliodorus beaten and Jacob wrestling, while that on the ceiling is a Saint Michael triumphing over Lucifer. I have not space for any adequate criticism of these works, but may say that the subjects suited the artist's genius exactly, and that he did himself justice. Whether art of that character, which is entirely wanting in repose, is suitable to mural painting, is another question. I think it is not. I believe that if the calm stability of architecture is to be happily accompanied by painting, the pictorial accompaniment should neither be too active nor too loud. It ought to be serene, calm, majestic, and severely conventional. In a movable picture the artist may display as much fire and impetuosity as he pleases; if the owner afterwards hangs the work in a wrong place it is not the artist's mistake, and it is easily remedied: but mural painting becomes a fixed part of the edifice, and the feverish energy of Delacroix seems out of harmony with the stately and massive architecture of St. Sulpice.

X.

PARKS AND GARDENS.

THE parks of London are so magnificent, so far superior to those of any other capital, that we Englishmen are naturally exposed to the mistake of measuring all other town parks by that standard, and then despising them accordingly. I say "mistake," because it is clearly an error to compare anything with a quite exceptional example of its kind. A man may be an admirable swimmer without being in any way comparable to the wonderful man who threw away his life at Niagara; a church may be a noble and interesting building without being half so large as the enormous cathedral at Rome; and a town park may be infinitely precious to the inhabitants of a great city, though it would look small on the banks of the Serpentine. A Londoner can never judge of town parks with any fairness if he is constantly thinking of his own. The right way to estimate such possessions is not the comparative, when comparison can lead to no result. If you wish to buy a book it is well for you to be told that there is a better and bigger work on the same subject, as perhaps you can afford to get it; but the Parisians cannot have Hyde

Park. They have their own places of recreation, and, especially during the last thirty years, there has been a laudable desire to multiply such places, and make them both prettier and more convenient; but there is no attempt to rival the parks of London.

Even if Parisian town-councillors had been disposed to make the necessary sacrifices, such parks would have been impossible in a city enclosed by fortifications. Let us remember what the history of Paris has always been. The town has always been a fortress; ring after ring of military wall has defended and limited it, nor was an old ring ever demolished until it had been made needless by the larger one outside of it. In the cramped interior of a mediaeval city the nearest approach to a park was simply a private garden, unless when land was enclosed, as it was within the wall of Philip Augustus, as a provision of building-land for future necessities. Such land was usually cultivated for profit until the time came for covering it with crowded houses and narrow streets. Unfortunately, too, it invariably happens that the value of open spaces is never strongly felt until the city has grown to a great size and has generally covered the land which would have been most convenient for a park. The existence of some of the most important open spaces in such cities is due to the merest chance. Some king or queen has had a fancy for a palace or a garden just outside the wall. A considerable space of land has been enclosed for that purpose, and so protected from miscellaneous buildings. Afterwards the old wall has been removed and a new one built at a distance, and then the land happened accidentally to find itself within the city. In future ages royalty prefers some other garden, or else royalty is abolished, and then the open space is preserved as a popular recreation-ground. That is one way in which a town park may come into existence; another way is very different from that. A space of ground may be out of the way for a long time, and so irregular as to be inconvenient to build upon. Afterwards, as the town spreads, this piece of awkward ground is surrounded by houses and becomes valuable. Then the question arises how to make it most useful, and the town or the Government turns it into a sort of park or garden. In all this there is very little real planning of open spaces for the best advantage of the public.

There was a time when the garden of the Tuileries lay just outside the wall of Paris, the *enceinte* of Charles V.; and now it happens exactly in the same way that the Bois de Boulogne lies just outside the present wall, and if a new belt of fortifications is made at some future time, the Bois de Boulogne will be within the city. So the space of land occupied by the park of the Buttes Chaumont lay outside of the fiscal wall erected under Louis XVI., but it was afterwards included within the fortifications of Thiers.

A short general account of the open spaces of Paris might be written as follows: The spaces of chief importance within the present walls are the gardens of the Tuileries and Luxembourg, the Champs Elysées,

Parks and Gardens.

the Champ de Mars, with the garden of the Trocadero opposite to it across the Pont d'Jena, the Jardin des Plantes, the Parc Monceau, and that of the Buttes Chaumont. It would seem out of place to mention the cemeteries here, but Parisian cemeteries are really little else than very large, well-kept gardens dedicated to the dead; and they are constantly visited by relatives and friends, so that, in fact, such great cemeteries as those of Mont Parnasse, Montmartre, and especially Père-la-Chaise, are places at least of retreat from the noise of the city, though the pleasure to be found in them belongs to the pleasures of melancholy. Just outside the present walls we have the Bois de Boulogne to the west and that of Vincennes to the east. Within the town there are now a considerable number of small gardens, with seats and fountains, besides trees, flowers, and a little space of lawn. These little gardens are always called "squares" by the Parisians; they have become immensely popular, and are most precious to the inhabitants of crowded streets at a distance from the Tuileries or the Luxembourg.

The origin of the Garden of the Tuileries is as follows: In the fourteenth century it was a region of market-gardens, brick-kilns, tile-kilns (whence the name), lime and plaster kilns, and potteries, interspersed with small summer residences for the citizens, at that time without the walls. It was in the highest degree improbable that such a region would be covered by anything better than a labyrinth of narrow streets; but it so happened that a large portion of the land fell into the hands of one family, called Le Gendre, about the beginning of the sixteenth century, which rendered it possible to make, from them, an important purchase at once. The Duchess of Angoulême, mother of Francis I., lived in a palace then existing, called *le palais des Tournelles*, and there was some horribly bad drainage near that dwelling, so that the most evil exhalations from a great mismanaged sewer offended the royal nostrils, and she looked out for a healthier and less odoriferous dwelling-place. There was a villa in the region of the Tuileries which sufficed for her purpose, and her son procured it for her, with a considerable estate of ground which belonged to the family of Le Gendre. At that time there was not the slightest intention of erecting a palace there; the Duchess simply wanted a summer residence for health's sake, and afterwards she lent it for life to Jean Tiercelin on his marriage. He was *maître d'hôtel* to the Dauphin.

This was the beginning, and the reader knows already what very much larger projects occupied the mind of Catherine de Medicis, who wanted an important palace, and built part of the Tuileries, which she hardly ever used. Her palace has already been described; her garden was an exceedingly formal affair, so that a map of it looks like the map of an ancient Roman city, with alleys always crossing each other at right angles. It was bounded to the north by a long riding-school, situated where the Rue de Rivoli is now, and to the west by a bastion close to the present Place de la Concorde.

Parks and Gardens. 179

The garden of Catherine de Medicis was in its perfection in the latter part of the sixteenth century, but it was altered to conform to later fashions. A century later the great principle of the right angle was abandoned, and both acute and obtuse angles, with segments of circles, were freely employed in the outlines of the beds, while their internal floral decoration was in flourishes of unrestrained curvature. In the latter half of the eighteenth century the flower-beds were restricted to a limited space in front of the palace, and beyond this were trees in plantations crossed by alleys at right angles. This arrangement has in the main been preserved to the present day, except that the flower-garden is now laid out differently; yet even here there is some respect for the old plan in the preservation of the basins and in the outline of the four sections westward of the great basin. The sections nearest the Louvre have been laid out afresh; the large octagonal basin near the Place de la Concorde remains exactly as it was a hundred years ago. Not one of the basins dates from the original garden of Catherine de Medicis.

The connection of cause and effect has seldom been more remarkable than in this instance. A bad smell which enters the palace of a royal lady in the sixteenth century is the reason why a great Republican city in the nineteenth has a garden for recreation precisely in the most convenient place. One special function of royalty in France appears to have been to prepare pleasant places for its heirs and successors, the people. It is well that the people know the value of such places.

The destruction of the Tuileries by the Communards was an exceptional act committed by a small minority in an hour of frenzied exasperation; the French people generally are fond of architecture and gardens, and proud to possess them. The garden of the Tuileries is likely to be preserved to a very remote future. At the present time it may be described as a sort of wood between two ornamental spaces. The trees in the wood (principally horse-chestnut and lime trees) make a noble avenue down the middle; but the ground beneath them is a desert trodden constantly by thousands, so that there is hardly room for a single blade of grass. At the east end of the garden the lawns are protected and kept in great perfection, as they are in all the public gardens of Paris. What the French call the *salles de verdure* of the Tuileries are, with their statues and the massive trees beyond, very beautiful examples of the classic taste in gardening.

When the lawns are only protected by low borders children are tolerated upon them. The garden of the Tuileries is the earthly paradise of Parisian childhood; and for any person of mature years who takes pleasure in watching the ways of children, a quiet seat there is an excellent post of observation. The extreme quickness and mobility of the French nature, and especially of the Parisian nature, are never better seen than in the children of the Tuileries. The wonder is that children can play so freely and happily when they are so fashionably dressed; the explanation must be, that as they are always dressed in that manner when out of doors

Parks and Gardens. 181

they live in a state of unconsciousness of fine clothes, which would be impossible in the country. The dressing of children is carried too far in all French towns; it seems as if they were little dolls for milliners to try expensive experiments upon. Any person who takes an interest in such matters has only to go and listen to

GRANDE ALLÉE DES TUILERIES.

a band on a sunny afternoon, when he will see a number of over-dressed little beings disporting themselves prettily enough.

The great defect of the Tuileries garden is the uninteresting nature of the ground itself, — a dead level, enclosed by straight lines. The terraces are an exception, though they also are straight, and seem to me wearisome; but this is merely a personal impression, and I know that there are many people who take a mysterious pleasure in walking on gravel flat as a barrack-yard between two monotonous rows of trees. What is really noble and remarkable in this garden is the frequent combination of sculpture and architecture with foliage, — a combination that never loses its charm, as the severe lines of stone and marble, and their gray or white color, excite in the eye a longing for the graceful masses of foliage and a desire for the priceless refreshment of its green. It is curious how little of a loss to the garden has been the destruction of the palace. Its removal has opened the magnificent perspective of Visconti's Louvre, which is quite sufficiently massive and imposing to fill up a distance effectively.[1]

The most complete contrast to the garden of the Tuileries is the Parc des Buttes Chaumont, situated

[1] It has been for some time proposed to erect a new palace of art on the site of the Tuileries, but the French Parliament has hitherto refused to sanction the plan. However, a Parisian friend tells me that M. Garnier, the well-known architect of the Opera, has prepared drawings for such an edifice, which is likely to be erected in course of time. There is evidently no intention of joining it to the pavilions de Flore and de Marsan, as they have new and magnificent fronts where such a juncture would have to be effected. The new palace, therefore, will probably be a completely isolated building, or else it may be connected with the pavilions by a light open screen in the form of an arcade. Whatever is done, it may be taken as certain that, with the present accumulated experience of the style, any modern Parisian architect of proved ability will produce a far better work than the old palace of the Tuileries ultimately became, and one much more in accordance with the buildings erected by Visconti, henceforth inevitably dominant over the whole edifice.

Parks and Gardens. 183

in a place that seems quite out of the way of visitors, though great numbers of them go there. It is near the northeastern corner of Paris, between the Boulevard de la Villette and the fortifications. There is a natural hill belonging to the high ground of Belleville, and this hill was partly cut up into quarries, chiefly plaster quarries, which left a broken and precipitous appearance,

LAC DES BUTTES CHAUMONT.

suggestive of great possibilities to the enterprise of a modern landscape-gardener. When this part of the city was laid out afresh in the year 1866, it was determined to reserve the roughest and most hilly portion of the ground as a pleasure-ground, greatly needed by that populous and unfashionable quarter. It is not very extensive, only sixty-one English acres; and this want of size is a serious defect, because one sees the surrounding houses too closely and too easily for the illusion of wild scenery to be possible; but it is very amusing and interesting to see with what extreme ingenuity the clever gardeners have made the most of their opportunity. By the help of a little willingness to be deluded, the visitor may imagine himself to be, — not in Scotland or Wales certainly, nor indeed in wild natural scenery anywhere, but in some picturesque park in Derbyshire; and to get so much of Nature as that is a great thing indeed in Paris. There is a pond, of course; but this pond excels most others in the possession of a precipitous rocky island, approached by a suspension-bridge from one shore, and by a lofty arch from another, and on the top of the island is a copy of the little temple of Vesta at Tivoli. Besides this, the land in the park rises to a considerable height in a steep green hill of pleasant shape with a wooded crown, and a rivulet makes music as it flows and falls happily from the wood to the lake. The water, no doubt, is real water, and the stones that it flows over are real stones, though placed there by human labor; neither is there any deception about the aquatic plants that grow gayly by its margin; but how comes the rivulet there? What is "The Stream's Secret"? Alas, for poetry! The secret in this case is a steadily toiling steam-engine on the banks of St. Martin's Canal, which persuades the water to go up the hill in a pipe, that it may come down again as we see. And now that I have told the stream's secret I will go yet a little further and tell mine, which is that

LE TROCADERO.

the poor little imitation rivulet seems to me distinctly and decidedly the pleasantest thing in Paris.

The park possesses a cave, which is impressive from its height, but wanting in the obscure depth of the great caverns, which inspires one with a sort of vague apprehension; and in the cave is another purling rivulet, so that the place is a paradise of shade and coolness in the sultry Parisian summer. From the temple, and also from several different places on the higher ground of the park, the views of Paris are very extensive; but they do not answer in all respects to its great reputation for beauty. It is true that in the remote distance you have hazy visions of towers and domes, and, as in all such city views, the sublimity of what seems an infinite world of houses; but you have also, in close proximity to the park itself, a region studded with long chimneys belonging to works of various kinds, and bearing a very close resemblance — I will not say to Liverpool or Manchester, for that would be an exaggeration — but at any rate to one of our minor manufacturing towns. The number of long chimneys in or near Paris has increased during recent years. Industry has made more visible progress than art, and there is some ground for the apprehension that in course of time the French capital may lose her beauty from this cause. The long chimneys interfere, even now, with the beauty of distant views. From the parapet near the Passy stairs I counted sixty-three of them this year, looking down the Seine and a little to the left. To a visitor from the north of England they are a reminder of home; but as English chimneys are equally tall, and emit smoke not less abundantly, why travel so far southward to see others of the same kind? The French are rather proud of them; their artists paint them in big pictures of the Seine, their *industriels* have them engraved for their advertisements.

AVENUE DES CHAMPS ELYSÉES.

Of recent improvements in Paris, there is nothing prettier or more needed than the garden on the slope between the Palace of the Trocadero and the river. It has been extended by another garden on the opposite bank of the Seine, taken from the Champ de Mars, and which now seems a continuation of the Trocadero Garden joined by the Pont d'Jena. The Champ de Mars now ends in a sort of terrace with a balustrade;

and on a fine starlight night a visitor can hardly spend an hour in a manner more likely to be remembered afterwards than in quietly leaning on that balustrade, and giving himself up to the influences of the strange and wonderful scene. Behind him is the vast, open, desert space of the Champ de Mars, silent and empty as so much land in the Sahara, and yet which has been the theatre of so many historical spectacles. There is no place in the world where the contrast between past and present — between many different pasts and the one monotonous present — is so striking and decided. No place in the world presents such a perfect *tabula rasa*, unless it be some area of salt water where fleets have fought and tempests raged, and where to-day no sound or motion disturbs the summer calm. The garden of the Tuileries was the chief scene of the Festival of the Supreme Being when Robespierre made a speech full of piety and virtue, and burnt the effigies of Atheism, Ambition, Self-seeking, and False Simplicity. Yet that memorable festival was also celebrated on the Champ de Mars; and on the next great occasion, the Festival of Federation, the whole ceremony took place there in the presence of three hundred thousand spectators, who stood upon embankments laboriously raised on purpose. Stood! nay, they sang and danced, in an ebullition of patriotic happiness. There was an altar in the middle, — *autel de la patrie;* and there was a throne near the military school, whereon sat poor Louis XVI., whose head still preserved its connection with his body. Talleyrand said mass, Lafayette rode about on a white horse. There was a great deal of solemn taking of oaths, in which the King and the President of the Assembly took part. After this, we learn that the *fédérés*, while they stayed in Paris, displayed a sincere enthusiasm for the king, the queen, the little dauphin, the constitution, and the Assembly.

AU JARDIN DU LUXEMBOURG.

In 1815 the desert of the Champ de Mars was covered with another crowd; there was an altar once again, with an officiating prelate, and a throne with another sovereign. It was now the *Champ de Mai*, though the ceremony took place on the 1st of June, — that fateful month which was to contain the date of Waterloo.

Parks and Gardens. 189

Napoleon came in coronation state, with a silken coat, a feathered cap, and the imperial mantle, in a state coach drawn by eight horses. Like Louis XVI., he, too, sat upon a throne, and received homage, and gazed over an ocean of human beings. Thiers says that almost the whole population of Paris was in the Champ de Mars that day; and it is certain that there were fifty thousand soldiers and a hundred pieces of artillery. It was the last imperial ceremony of the First Empire. When Napoleon laid aside the imperial mantle that day, as he left the throne to distribute colors, he had done forever with imperial state. Nothing remained for him but a fortnight of rough life as a soldier, to be followed by a crushing defeat, a wretched exile, and a miserable death.

There has been no public ceremony in more recent times so memorable as the Champ de Mai, but many of us remember the military reviews of the Second Empire, which were very striking spectacles of their kind; and then came the great exhibitions with their enormous buildings, which have vanished like enchanted palaces in fairy tales. Changes which in other parts of Paris have required centuries are effected in a year or two on the Champ de Mars. Its permanent condition is that of perfect emptiness and aridity, but occasionally it is the scene of wonderful concentrations of humanity. The exhibition of 1867 is like a page of ancient history already. How remote it seems! I remember, as if it were yesterday, the Emperor arriving on the opening day, accompanied by his wife and child, and looking neither well nor happy. Coming events were already casting their shadows. A German waiter calmly told me that there would soon be a war between France and Prussia, and he looked forward to the result with confidence. The Empire was already tottering, nobody counted upon its long continuance. When the next great exhibition palace had been erected in 1878 the object of the display was a revival of cheerful energy after dispiriting disaster. It was a far more imposing structure than the first, and surrounded by quite a town of buildings filled with the densest crowds. Now, again, the Champ de Mars is a *tabula rasa*, and all that is to be seen across its vast expense at night is perhaps the lamp of a solitary cab crossing near the École Militaire, and proving the distance by the excessive apparent slowness of its motion.

The Trocadero Palace, which is left as a permanent legacy of the Exhibition of 1878, has often been severely criticised on account of its large protuberant central body and its great curving arms. French people say it is like Victor Hugo's "Pieuvre," but these criticisms can only be applicable when the building is seen from a little distance. From the Champ de Mars it presents a most imposing appearance, especially on a fine night. The site is incomparable. The whole width of the building has a clear space before it for nearly fifteen hundred mètres, and it stands upon a stately height, from which a beautiful garden slopes down, at first rapidly, then more gently, to the river, crossed there by one of the finest of its bridges; then comes

LAC DU BOIS DE BOULOGNE.

another wide space of garden, and beyond that the Champ de Mars. ˙ When the sky is full of stars and all this scene covered with lights like an illumination, it is enough to inspire a poet, and would in itself be in the highest degree poetical if it were not so modern and so easily accessible. Only forget that it is in familiar Paris, a day's journey from London, forget that these are gaslights, imagine that those stately domes, those lofty towers, are the dwelling of some mighty and mysterious Oriental potentate, and by getting rid of the obtrusive commonplace and familiar, you may enjoy the real magnificence of the scene. On one occasion, the National Festival of 1883, especial art was employed to enhance the beauty of the spectacle, and then it reached a degree of splendor that no Eastern sovereign ever attempted.

The French have a great liking for open and extensive city views. If London belonged to them, they would clear away all the buildings between the British Museum and Oxford Street, if they did not carry a broad avenue down to the Strand. The feeling of openness in Paris is immensely enhanced by the way in which several different spaces are often happily combined. A man's garden gains in apparent liberty by the width of his neighbor's field. The garden of the Tuileries has the Place de la Concorde and then the Champs Elysées, with the long and broad avenue beyond, up to the triumphal arch. There is a general feeling of openness about the Seine, with the Champs Elysées on one hand and the Esplanade des Invalides on the other. As for the Elysian Fields themselves, they need no detailed description. They do not seem to be very much of an Elysium, but they offer shade

LA NAUMACHIE, — PARC DE MONCEAU.

and seats and cool draughts of Vienna beer. The word "fields" is too ambitious. There is nothing here but a little wood with tidy walks, and grass kept green by perpetual spray, — altogether a pleasant small substitute for real nature, like the rivulet fed by the steam-engine.

Parks and Gardens.　193

The Palais de l'Industrie here is better named perhaps than if it had been more ambitiously entitled a Palace of Art, since the pictures at the annual Salons are chiefly industrial products on an extensive scale. The crudity of color which used to be the peculiar distinction of inexperienced English painting has of late years been attained, or surpassed, by a multitude of energetic Frenchmen; and as they combine with it a national delight in self-assertion and a peculiar enjoyment of the horrible, the present Salons are not by any means scenes of unmixed or refined pleasure, though held in the Elysian Fields.

The garden of the Luxembourg is one of the most frequented places of recreation in Paris, and it is much to be regretted that in the latter days of the Empire it was diminished by cutting off a large acute-angled triangle at the upper end of the *pépinière*, to make room for the Rue de l'Abbé de l'Épée and other streets. Some important buildings, including the École des Mines, the Pharmacie Centrale des Hôpitaux, and a large new Lycée, have been erected on ground that formerly belonged to the nursery or the garden of the Luxembourg, and this at a time when the rapid increase of Paris in every direction made it more than ever desirable to preserve all open spaces with the most jealous care. It was a piece of economy, and of very unpopular economy, the only practical reason in its favor being that the new Rue de l'Abbé de l'Épée rendered communication a little easier. In the remaining ground there are five pretty gardens with lawns and a considerable number of parallelograms planted with trees; and these, with the more or less open spaces between them, serve as playgrounds for the children. The eastern side of the garden is the favorite resting-place for grown-up people, who sit there on many hundreds of chairs. What I have called especially the gardens are spaces laid out as lawns, with winding walks, a sufficiency of trees for shade, and plenty of garden-seats. The lovers of tranquillity seek these retreats, and sit quietly watching the fine spray that spurts from the water-pipes on the lawn and makes little rainbows over the grass. There are landscape-painters who have studios in that quarter and who prize these little gardens, not as if they were wild nature, but for the degree of refreshment they afford to eyes weary of walls and pavements.

The woods of Boulogne and Vincennes both lie immediately outside the fortifications, and are good specimens of what the French understand by pleasure-grounds. Both have artificial lakes of considerable size with islands, and the woods are pierced in various directions by well-kept roads. Although the recreation-grounds within the walls of Paris are much smaller than the London Parks, the Bois de Boulogne is very much larger. Its area considerably exceeds two thousand acres, which is much more than that of all the London parks put together, and it includes about sixty miles of rides and drives. Almost every reader of these pages will be aware already that the Bois de Boulogne is the resort of all Parisians who can afford to keep carriages and horses; and it is visited on holidays by many thousands of the middle and working classes. I heartily appreciate the wisdom of setting apart a great space of land for public recreation, the noise and crowding of city life make such places necessary, and if they were not firmly protected now the future would be entirely deprived of them; but I cannot say that the Bois de Boulogne has ever seemed to me delightful. Any country lane that winds about among fields, and crosses a stream here and there, now hiding itself in a dell, now affording a view from a little eminence, suits my taste far better than well-kept carriage-drives between dense, monotonous groves of green. The Bois de Boulogne is one of those places in which a lover of real landscape feels himself to be most a prisoner. The very perfection with which it is all kept is enough to make him long for a little uncared-for nature. It is difficult to imagine any more tiresome form of recreation than that of a wealthy Frenchman, who has himself dragged along those miles and miles of road past millions of trees that always seem the same. The real amusement of such a Frenchman is to criticise people and equipages; but he might enjoy equal facilities for such a mental occupation on a chair in the Champs Elysées.

The prettiest public garden in Paris is the Parc Monceau, not to be in any way confounded with what we call a park in England, yet a piece of ground very tastefully laid out with undulating lawns, shady trees, statues, and a little sheet of water, that reflects a Corinthian colonnade in a half-circle. Nothing can be more elegant than this colonnade, which has been preserved from the times of the early French Renaissance, but nobody knows exactly from what palace or monument it was taken. In its present situation it seems like a remnant of antique architecture in some graceful picture by Claude, and one is grateful for the good sense that has saved it from destruction. Lalanne once made a very poetical charcoal drawing of it, which has been reproduced in the series of his charcoals. This is one example the more of the happy combination of architecture with foliage and water. Set up in the British Museum, these columns would signify comparatively little; but with graceful foliage and a mirror of water, they are charming.

XI.

MODERN PARISIAN ARCHITECTURE.

OF all modern cities Paris is the one in which the notion of architecture is most generally prevalent. In London, as in all our English towns, the ordinary builders have worked without any notion of architecture at all, and the real architect has seldom been called in unless to erect some important public building. In Paris architecture of some kind is very common. Thousands of houses have been erected with a definite architectural intention; and this architectural tendency has of late years become so habitual that in the better quarters of the city a building hardly ever rises from the ground unless it has been designed by some architect who knows what art is, and endeavors to apply it to little things as well as great.

Modern Parisian architecture has settled definitely into a new form of Renaissance. I find it convenient to separate the early elegant Renaissance (of which there are still some charming examples in France, full of graceful art and invention, combined with delicate finish in workmanship) from the heavy, ascetic Renaissance that followed it, in which there was no enjoyment, no fancy, no delicacy, no imagination, and scarcely a trace of any other feeling than pure pride in size, and cost, and heaviness. The Hôtel de Ville and the Court of the Louvre belong to the elegant Renaissance. St. Eustache is an attempt to marry that Renaissance with Gothic, but the west front of St. Eustache is in that tiresome style which in my own mind I always think of as the stupid eighteenth-century Renaissance. Now the effort of modern French domestic architects has been to start afresh with a second elegant Renaissance, and in a great measure they have succeeded. They have emancipated themselves from the dulness and heaviness of their immediate predecessors; they have allowed themselves some variety, some free play of the fancy and intelligence; and although their art is seldom strikingly imaginative, it is full of interesting experiments. A firmly prejudiced visitor from another country might easily shut his eyes against it altogether, and say that it is all exactly alike, because it is generally governed by the prevailing taste of the time; but the real interest of it consists in the variety that underlies a general fashion. The fashion is a cheerful and free Renaissance; the variety consists in the use of as much freedom as is compatible with a dominant idea.

A few experiments have been tried with mediaeval forms, or with mediaevalism passing into Renaissance; and one of the most successful of these latter is the building of the Historical Society in the Boulevard St. Germain; but true Gothic has been definitively and wisely abandoned. It has been wisely abandoned because the pointed window-head never looks its best unless there is either a gable or a larger Gothic arch above it. A Gothic window does not look well in a room with a flat ceiling, and a row of Gothic windows do not look in their right place under a long straight cornice, like those in a modern street. Under the gables of a mediaeval street they might look better, but a row of gables, like the teeth of a saw, is neither the most rational nor the most economical form of roofing for street houses, and it has been finally and completely abandoned. You may, it is true, fill up your Gothic window-head with a tympanum in the shape of an inverted shield, and so get a square head for the real window inside, but such a process is unnecessarily expensive. Evidently the plain course was to adopt the straight head, the simple horizontal stone of classic architecture, and that settled the question in favor of Renaissance forms. The condition of another art may also have had its influence. Modern French sculpture comes almost directly from antiquity; it has come from Greece and Rome through the Renaissance; it has not come out of Gothic forms by evolution. Modern French sculptors can be trained to do something that will pass with unobservant people as a substitute for Gothic sculpture, but it is not natural to them. They try to make their work *naïf*, but they only succeed in making it stiff; they have not the true Gothic *naïveté*, and they cannot have it; they cannot have that delightful blending of pre-scientific simplicity with deep feeling and shrewd observation which characterized Gothic art. They know far too much, and when they feel, they do not feel in that manner. Now there are great numbers of sculptors in Paris who have received a considerable amount of artistic instruction, but who cannot keep themselves by making statues that only the Government buys, so these men turn their talents to ornamental sculpture. Their education in art has been wholly classical, and their practical influence upon modern architecture has been very considerable, because the architects know exactly what sort of ornamental work the carvers are fit to do. In short, the sort of domestic architecture that naturally springs from the Parisian mind, such as education has fashioned it, must be a form of Renaissance architecture, and none other. A literary critic has remarked that we are much nearer, intellectually, to the classic authors than to the mediaeval ones; and it is not less true that the architects and workmen of modern Paris work in Renaissance forms as naturally, and when left to themselves as inevitably, as they speak French. Such forms have no longer anything of an imported style; they seem as much a product of the soil as if they had been invented by the ancient Gauls.[1]

[1] I remember trying, many years ago, to get an oak pedestal carved in Paris. It was supported by three griffins, and I had drawn Gothic griffins, but the carvers I applied to immediately made sketches of Renaissance griffins, and said they would do much better. As that was the transformation I had been most anxious to avoid (for the particular piece of furniture in question), I gave up the project. The carvers were highly intelligent workmen, yet quite incapable of conceiving anything that was not in a Renaissance spirit. I had another example of the same difficulty afterwards. A French draughtsman was employed to copy with the pen, for photographic reproduction, a series of pictures by an English pre-Raphaelite artist. In making the copies he eliminated all the pre-Raphaelite characteristics of feeling and style, and substituted those of

Any adequate account of modern architecture in Paris would require a volume to itself, and such an account could not be made interesting or intelligible without the help of minute and abundant architectural engraving, while it would find few readers outside the special public that really studies architectural subjects. All that can be done here is to give a general account of prevailing tendencies. The reader who cares to follow out the subject may do so with the help of the works issued by the Parisian architects themselves.

The mediaeval arrangement was to turn the gable towards the street, and in a mediaeval city every house had its own gable, whence the old French expression concerning a well-to-do citizen that he had *pignon sur rue*. Nothing strikes us more in the old engravings of Paris than the wonderful number of gables, especially round such open spaces as the Place de Grève and the Cimetière des Innocents. Many of these survived until the eighteenth century, but they belong essentially to Gothic times. The greatest clearing away of gables appears to have taken place in the seventeenth century, after having been begun a hundred years earlier or more. Under Louis XIV. every house-builder appears to have turned the eaves towards the street like the architects of the present day; and as in succeeding reigns the old houses were finally removed from the

the Renaissance, thereby, of course, entirely falsifying the intentions of the original painter. I believe he did this quite unconsciously; at any rate, he was evidently incapable of supposing that the peculiar interest of the originals lay precisely in those very characteristics that he eliminated.

bridges and quays the eyes of the citizens became more and more accustomed to continuity of line.

Still, although the eaves were turned towards the street, the gable was not entirely abolished, because it occurred at the end of every row of houses. Instead of being innumerable, the gables had become few, but that was the extent of the change. Now in modern Paris the gable is entirely abolished except in a few private mansions where the owner has followed his own taste; and the abolition of the gable is one of the most important of all decisive changes. It cuts modern architecture completely adrift from mediaeval. And please observe that this revolution has not been accomplished, as in London, by the abolition of the visible roof. There are plenty of streets in London where you cannot see the roofs of the opposite houses. In Paris it is not so. There the roof is rightly felt to be of the greatest expressional importance; but instead of ending with a gable, it is truncated either with a roof sloping at the same angle as the other, or with a curve when the rest of the roof is arched. The value of space in Parisian houses has led to the very general adoption of arched or bulging roofs, which have the advantage of allowing so much more head-room, a truth well known to all who use tents and wagons. In cases where the curve is not employed, the roof often begins by being exceedingly steep and then comes to an angle from which it slopes back rapidly to the ridge, and in the steep part of it there is a row of dormer windows.

The modern Parisian house, then, is characterized by a visible roof, curved or angular, with dormer windows in it, but not any gable either towards the street or at the end. The windows are flat-headed, they are very frequently provided with an entablature and with lateral mouldings, while in a great number of the better class of houses the stonework that surrounds the window is carved more or less elaborately, but almost always with knowledge and good taste. Great use has been made of balconies as an element of architectural interest and an excuse for tasteful decoration. They are always supported on massive stone brackets which in every instance show at least an attempt at design, while many of them are beautiful in form and enriched with excellent ornamental sculpture. The doorways, in modern houses, are generally of importance. The French habit of living on flats makes one doorway the entrance to many dwellings, so that an amount of ornament may be lavished upon it which would be extravagant and impossible for a single tenant. The finest of such doorways consist of a lofty stone arch decorated with sculpture and filled with a tympanum of oak with folding-doors below, large enough for the passage of carriages. The woodwork is thoroughly sound and well finished, very strong and massive, and left almost of its natural color, but varnished. Carving is employed on the woodwork, but generally in moderation, and always in perfect keeping with the stone-carving on the rest of the edifice. There is also a taste for massive handles plated with nickel or silver and set in small slabs of marble. Up in the panels of the tympanum there is often a window belonging to the *entresol;* and when this occurs

DOORWAY OF A MODERN HOUSE.

the surroundings of the window in mouldings, carvings, and panels are as carefully designed, though in woodwork, as the masonry of the house itself. In such a house there is not an inch of surface from roof to basement that is not ruled by thoughtful care and taste, accompanied by sufficient knowledge. I do not speak of genius and inspiration, these are as rare in architectural as in literary work; but it is a great thing to have banished ignorance and bad taste. It is a great thing, too, that house-builders should have got well out of that negative condition of perfect dulness, of incapacity to desire or apprehend the beautiful, which produced such houses as those in Harley Street. Even in Paris itself, although the builders from Louis XIV. to Louis Napoleon sometimes erected interesting separate mansions, they treated houses in rows with wearisome monotony whenever they had power to build a row of houses at all. The last houses on the Pont au Change, which were finished in 1647 and demolished in 1788, were as dull as domestic architecture of the last century in London. The supplementary buildings of the Hôtel Dieu on the left bank of the Seine, which were completed in the eighteenth century, had not more architecture than a cotton-mill, and the houses behind them were no better. The pretty modern Parisian house does not date farther back than Louis Napoleon, and at first it was monotonously repeated. The desire for variety came in due course, but it was only towards the close of the reign that the possibilities of the new style came to be thoroughly understood. It still requires, I think, a more obvious and clearly visible variety, though it is easy to fall into the common error of not observing the degree of variety that there is. More will have to be said about street architecture in the next chapter. For the present I desire to point out a peculiar effect of the increased attention paid to the architecture of houses. Since so many of the houses have been made lofty and beautiful, many new public buildings have been very strongly influenced by them, both as to their proportions and their style of architecture. Some readers will remember an absurdly small church at Geneva with a miniature tower, which is surrounded by very lofty modern houses. In a village of low cottages such a church would look respectable; at Geneva it is like a model set there for the people to look down upon from their windows. On the contrary, Rouen Cathedral *gains* by contrast with the old-fashioned houses close to it, which are not on a great scale. The merit of Parisian architects is to have perceived the new necessities in public buildings created by streets of magnificent private dwellings. If the ordinary architecture of a city is on a large scale and richly decorated, its public buildings must still distinguish themselves by greater richness. One consequence of the reconstruction of Parisian dwellings has been the rebuilding, in whole or in part, of almost all those theatres that happened to be near new streets or squares. The Théâtre Français had a new front; the Opera was rebuilt with unparalleled magnificence; the Vaudeville had a narrow but strikingly rich curved *façade* at the corner of the Chaussée d'Antin, with Corinthian columns and Caryatides and a *fronton* crowned with a statue of Apollo. The new Théâtre de la Renaissance is a heavy but sumptuous

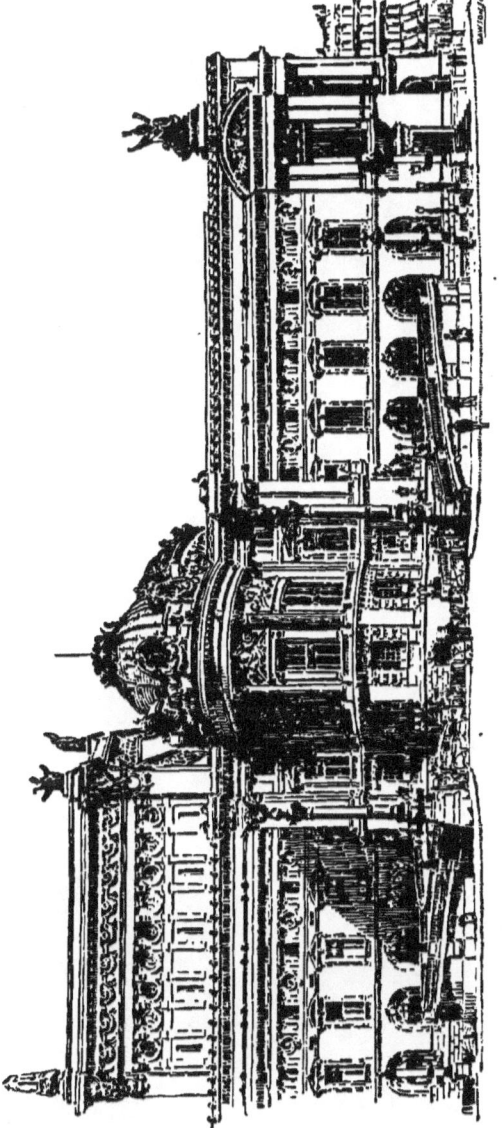

THE OPERA. SIDE VIEW.

structure, also adorned with Caryatides and Corinthian columns. The Gaîté was rebuilt in 1861 with a pretty arcade on marble columns in front of its open loggia. The Châtelet was built at the same date, and has also its loggia, but with statues under the five arches. The neighboring Théâtre Historique, which used to be the Lyrique, was also built under Louis Napoleon, though it has been rebuilt since in consequence of incendiarism by the Communards. The construction of these buildings, and of many others, was made a necessity by the handsome new houses. The Odéon belongs to the beginning of this century and is a plain, respectable structure. It may remain as it is because the houses near it are plain, old-fashioned dwellings of the same or an earlier date; but if the Odéon could be placed where the Opera is now, it would be too simple for such a situation.

Yes, the French understand the effect of neighborhood in architecture, an effect which may either completely destroy or wonderfully enhance the charm and interest of a building. I wish it could be said that the English understood this equally well, or were equally ready to make the sacrifices that are necessary to protect a building from being injured by its neighbors. The French are not always careful enough, or, at least, not always successful, as we see in the injury inflicted by large buildings on Notre Dame and the Sainte Chapelle; still the principle is understood in Paris, and very few public buildings of any consequence are inadequate to the situations which they occupy.

The most magnificent of recent structures, and one of the most happily situated, is the Opera. The situation has been created for it purposely. The front might have looked merely across a street, but a new street of great length was opened, that it might be seen from a distance. Besides this, arrangements were made for the convergence of several other new streets in front of the Opera, so as to give to its site the utmost possible importance. As the houses in these streets are all of them lofty and many of them magnificent, the Opera itself required both size and richness to hold its own in a situation that would have been dangerous to a feeble or even a modest architectural performance. The Opera was compelled to assert itself strongly, and if it had merits they must be of a showy and visible kind, — rather those of the sunflower than those of the lily of the valley. There can be no question that M. Garnier aimed at the right kind of merit, — showy magnificence, — but there are two opposite opinions about his taste. Like all important contemporary efforts, the Opera has its ardent admirers and its pitiless critics. Let me tell a short anecdote about this building, which may help us in some measure to arrive at a just opinion. Shortly after its completion several distinguished men, who were not architects, met at a Parisian dinner-table, and they criticised M. Garnier with great severity. Among them was a provincial architect, who remained silent till the others appealed to him. Then he said: " Gentlemen, when an architect undertakes to erect a comparatively small building it is still a very complex affair;

THE OPERA. THE PRINCIPAL FRONT.

and how much more so must be such a gigantic work as the Opera, where a thousand matters of detail and necessity have to be provided for, all of which the architect has to carry in his mind together, and to reconcile with the exigencies of art! Such a task is one of the heaviest and longest strains that can be imposed upon the mind of man; and if the architect does not satisfy every one, it may be because other people are not aware of the extreme complexity of the problem." For me, I confess that I know really nothing about theatres, except that they have mysterious difficulties of their own. I like being outside better than inside them, because to be outside is at the same time cooler and cheaper; and all I know about their peculiar form is that they generally have a gabled superstructure, which must be an awkward thing for an architect, and is in some way connected with scene-shifting. I humbly confess that the Parisian Opera seems to me a very odd sort of structure when seen from behind, and perhaps it might have been better to hide those parts of it. Yet I like to see the whole of an edifice, the complete work of the architect, and not merely a fine front, like the front of a shop. The truncated angles at the back have a decidedly weakening effect upon the design, but the corners were cut off in order that there might be an apparent correspondence between the building and the Rues Scribe and Glück. The rotundas on the east and west sides have a good effect in breaking their monotonous length, and their domes make a good accompaniment to the great flattened dome over the house.

The principle followed everywhere has been to conform the exterior to the uses of the edifice, which is right. The exterior dissimulates nothing, and consequently it looks like nothing in the world but what it is, — a great theatre; whereas the Vaudeville might be taken for the entrance to a bank, and the Odéon for a scientific lecture-hall and museum.

Whatever may be thought of the back and sides of the Opera, the principal front may be admired without reserve. The basement is a massive wall, finished plainly, and pierced with seven round arches. In the intervals between five of these arches are statues and medallions; on each side of the two exterior ones are groups representing Music, Lyrical Poetry, the Lyrical Drama, and the Dance. The contrast here of extreme architectural simplicity with figure-sculpture is excellent. Above is a colonnade of coupled Corinthian columns supporting an entablature, and between each two pairs of columns is an open space, in which a lower and smaller entablature, with a wall above it, is supported on smaller columns of marble. This wall is pierced in each interval with a circular opening containing the gilded bronze bust of a great musician. Above the great entablature, and immediately over each pair of coupled columns, is a medallion with supporters, and above each open space of the loggia is an oblong panel with sculpture. Then you come to the dome of the house and the gable of the structure above the stage. The effect of the whole is a combination of splendor with strength and durability. The use of

INTERIOR OF THE CHURCH OF ST. AUGUSTINE.

sculpture has been happy, and the sculpture has not been killed by the architecture, as it often is. On the other hand, it has lightened the appearance of the architecture, especially on the top of the edifice where the colossal winged figures are most valuable, — and so is that on the apex which holds up the lyre with both hands.

With regard to the interior, my humble opinion — the opinion of one who knows nothing about theatres — is, that the business of plotting for splendor has been considerably overdone. The *foyer* is palatial, but it is overcharged with heavy ornament, like the palace of some lavish but vulgar king. As for poor Paul Baudry's paintings on the ceiling, which cost him such an infinity of labor and pains, it does not in the least signify what he painted or how long it will last, for nobody can see his work in its present situation. There can hardly be any more deplorable waste of industry and knowledge than to devote it to the painting of ceilings that we cannot look at without pains in the neck, and cannot see properly when we do look at them. The grand staircase is more decidedly a success than the *foyer*. It almost overpowers us by its splendor; it is full of dazzling light; it conveys a strong sense of height, space, openness; it comes on the sight as a burst of brilliant and triumphant music on the ear. The mind has its own satisfaction in a work that is splendid without false pretension. All the materials are really what they seem. The thirty columns are monoliths of marble, every step is of white Italian marble, the hand-rail of onyx, sup-

THE CHURCH OF ST. AUGUSTINE.

ported by balusters of *rouge antique*, on a base of green marble from Sweden. We may admire the grand staircase or object to it, but it is honest work throughout, and may last a thousand years. The architect evidently took pride in it, as he has so planned the design that visitors may look down from galleries on four different stories all round the building. The house itself is much less original, with its decoration of red and gold, and the customary arrangements for the audience.

House architecture in the modern streets of Paris has led the architects to attempt the solution of a very difficult problem. They have endeavored, — I will not say to invent a new style of ecclesiastical architecture (for a really new style is not possible), but to adapt an old style in such a manner as to make it harmonize with the secular and domestic architecture of our own time. If the reader will glance back in memory at the styles of church architecture that have prevailed or been experimented upon since the beginning of French civilization, he will soon perceive that there really is not one of them that would not look isolated in a modern boulevard. The Romanesque and Gothic styles, in all their varieties, look completely isolated. A classic temple like the Madeleine looks out of place for various reasons, especially for its want of height in comparison with modern houses, and its prison-like absence of openings, so different from the modern wall, pierced with many windows. The architect of St. Eustache made a most important experiment in the union of Gothic principles with the details of the elegant Renaissance, but his example has not been followed. As for the dull and heavy architecture that I have ventured to call plainly the stupid Renaissance, it would look uglier than ever if placed in the neighborhood of intelligent and inventive modern domestic building. Contempo-

INTERIOR OF THE CHURCH OF LA TRINITÉ.

rary Parisian architects have endeavored to solve the problem by a free expansion of Byzantine ideas. The most interesting of these experiments is the Church of St. Augustine, where stone and cast-iron have been employed together. The use of cast-iron has been almost entirely confined to the roof and dome. The nave is crossed by light iron arches, with spandrels of the same material; and from these arches a metal column comes down to the ground on each side, set against the stone pier like a pilaster. These iron arches carry depressed vaults corresponding to the bays. The dome is almost entirely metallic; its ribs, and even the mullions of its windows, are of iron. Each bay of the nave consists of a round arch with three minor arches above it in the triforium gallery, themselves included in a higher arch, and in front of the triforium runs a gilded railing. The windows of the clerestory are round-arched, each with two lights, containing, in painted glass, figures of bishops or other ecclesiastics. Ornament is used in moderation, and is not in itself of an elaborate kind, a small lozenge-shaped or square panel being considered enough to vary a space of plain wall. The tympanum between the three small arches and the large one that includes them is richer.

The interior of the church has certain merits in a very high degree. It is not only spacious and airy in reality, but it looks so. I have seldom found myself beneath a dome that seemed so light and lofty. The nave is of great width, but there are no aisles, and the lateral chapels are unfortunate in shape, owing to the site, which compelled M. Baltard, the architect, to make them narrower and narrower as they approach the principal entrance. The exterior has been considerably injured by this necessity, which gives the whole edifice the appearance of being huddled together. The towers are too close to the dome, and the front seems to require lateral support of some kind. Much of the interest of this church as a piece of work consists in the difficulty of the site. That may have been one of the reasons for the employment of iron in the roof, as it caused so much less outward thrust, and the building could not spread itself laterally. Whatever the reason, the iron has been skilfully used, and in that respect, as well as in the character of all the other arts employed, this church is thoroughly of our own time.

The Church of La Trinité is another important example of modernism. The nave is very wide, and vaulted in a large round arch. The aisles are very narrow, and separated from the nave by an arcade of round arches supported on marble pillars between the piers. Above the aisles runs a lofty gallery, with a similar arcade and a pierced parapet. The space above the high altar is narrowed by the projection of two arcades, equal in height to those of the aisles, and finished by a continuation of the parapet just mentioned. There is a small apse behind. The church has a tower crowned with an octagonal dome and lantern, and flanked by two small lanterns, also with domes. The round arch is dominant everywhere, except over the niches and doors, where pediments have been frequently employed. In the perfect finish of the workmanship, the richness and excellence of the materials, and a general air of palatial

THE CHURCH OF LA TRINITÉ.

elegance, this church is quite modern and Parisian. It is curious to observe how well it holds its place between two large blocks of houses built at the same date with itself, which have round arches over the *entresol*, and louvre windows not much unlike the upper niches, and pilasters recalling those at the angles of the church. Opposite the Trinité, the houses at the angles of the Chaussée d'Antin are finished with domes. The balustrade immediately in front of the Trinité (behind the three fountains) is carried round the pretty garden, which seems in this way to belong to the church. For all these reasons this piece of ecclesiastical architecture is allied to its surroundings just as the Gothic cathedrals were in the Middle Ages.

Another example of the same modern style, founded upon the combination of the round arch with the classic capital, is the Church of St. François Xavier, near the Invalides. This church is plainer and simpler than the Trinité, and much smaller than St. Augustine. It is not in any way imposing, but it is interesting as one of the most honest attempts of a modern architect to build in a modern way. Such work is far less unsatisfactory than a thin attempt at Gothic like St. Clotilde.

In Paris, where there is really a modern style of domestic architecture, it is possible that in the future a corresponding ecclesiastical architecture may become habitual. Gothic is too remote from modern habits of design and too much isolated in the midst of modern houses. A heavy Renaissance like that of St. Sulpice is too much wanting in grace and cheerfulness. What really suits modern Paris is a sort of Renaissance, very delicate in workmanship everywhere, and combining readily with intelligent painting and sculpture. It should employ beautiful materials, fine marbles, gilded bronze, and other good modern metal-work. At St. Augustine the doors are in electrotype copper. Above all, the modern style should leave great liberty to the taste and fancy of each individual architect, because it is only in this way that any boldness of experiment can be possible, or any new ideas evolved. These modern churches show signs of real vitality, and in this respect are more hopeful than any mere *pastiche* of Gothic or Italian art.

XII.

THE STREETS.

THE English have invented the *house*, the French have invented the *street*. By this I do not venture to affirm or undertake to maintain that nobody lived in what were called houses before the existence of Englishmen, nor that ancient cities had not streets of some kind; but I mean that the English are the first people who have thoroughly understood the house and realized it, setting in this respect an example to other nations, and that the French are the first people who have thoroughly understood the street and realized a conception of it which has become a model of excellence in its own kind.

An Englishman who finds himself in some great Parisian street quite of our own time, such as the Boulevard Haussmann or the Boulevard Malesherbes, has nothing to do but simply confess that here is the ideal street,[1] and that his own Piccadilly and Oxford Street

[1] For our part of Europe and other temperate climates. It appears that narrow, tortuous streets, with overhanging stories and a space above that can be easily covered with an awning, are much preferable in hot countries.

are not yet the ideal. A street should not only be wide, for the facility of traffic, but it should be of the same width throughout, that there may be no local obstruction. The causeways for foot-passengers ought to be wide also, and there ought to be seats where they may rest when weary. Trees are not an absolute necessity, but next to space, air, and light, they are the greatest of all luxuries, not only for their shade, but for the delightful refreshment afforded by the green of their foliage in a wilderness of stone and mortar. With the blue sky and the passing clouds above, and the fresh green leaves on the trees, it seems as if nature were not quite banished yet.

True lovers of Paris (I am simply an admirer, and have no sentiment of affection for the place) take a keen delight in those broad *trottoirs* of the Boulevards. They walk upon them for the mere pleasure of being there, till absolute weariness compels them to sit down before a café; and when the feelings of exhaustion are over, they rise to tire themselves again, like a girl at a ball. They tell one that the mere sensation of the Parisian asphaltum under the feet is an excitement itself, so that when aided by "little glasses" in the moments of rest at the cafés, it must be positively intoxicating. These true lovers of Paris are most enchanted with those parts of the Boulevards where the crowd is always so dense that all freedom of motion is impossible; where half the foot-way is occupied by thousands of café chairs and the other half by a closely packed multitude of loungers. The favorite places appear to be the Boulevard des Italiens and the Boulevard Montmartre. The shops are, in fact, a great permanent exhibition of industry and the fine arts, wonderfully lighted at night,[1] and very attractive to those who visit Paris on rare occasions; but it is surprising how much of the illusion disappears after close and old acquaintance. You find the same things repeated, either identically or with slight changes that are easily seen through. It may not be exactly the same picture that you saw in the dealer's window three years ago, but very likely it will be the same kind of picture, set off in the same way by an enormously disproportionate frame on a background of dark red velvet, the whole so lighted that the gilding flames across the street. The bronzes are not quite the same perhaps, yet it is difficult to believe them new. There is the old green caricature-bronze, some meagre-limbed Mephistopheles; there is the Barye or Cain animal bronze; there are the multitudes of coppery girls, as evidently daughters of Paris as if they were dressed, and dressed in the fashion. Amid the glittering shops, where the object is to vanquish the eye by mere dazzle, you come upon the intensely respectable, excessively quiet shops, that invite only to repose. Nothing amuses me more in the French mind than its fine artistic faculty of taking up a *motive* and keeping to it. The faculty deserves hearty admiration, but the exercise of it is amusing because it is simply artistic, like acting, and has nothing to do

[1] On the old much-frequented Boulevards, but elsewhere early closing is beginning to prevail.

with character. You pass a glaring, frivolous-looking shop, full of gayety and glitter, and then you meet with a dark-looking, quiet shop, that looks like a retreat for a profoundly meditative mind, and is severely finished in ebony and stamped leather. What is admirable in such places is the determination to keep out the incongruous. It must be one of the keenest pleasures to plan a shop of the severe kind, to decide upon its sober colors, its rich yet simple decoration. I am not the inventor of the remark that the French have a genius for shop-keeping. Their love of neatness and order, their appreciation of pretty things, their talent for making the most of everything and showing it to the best advantage, all combine to make them masters in the art of managing a *devanture*. The proverb, *Marchandise bien parée est à moitié vendue*, is a piece of French mercantile wisdom. There are all varieties in the art of exhibiting goods. One dealer overwhelms you with quantity, but that is an appeal to the vulgar. The opposite policy seems far more refined and crafty. I confess to a sincere admiration for the tempter who displays very few but very exquisite things, and has the art of arranging them so that they shall help each other. One Parisian dealer in works of art showed very little, yet had a great collection. "You could fill a museum," I said, and was told that he did not consider it good policy to show many things at once.

The colossal shops that have sprung up in Paris of late years are beginning to employ architecture as an advertisement. The most curious instance of this is the

BOULEVARD ST. GERMAIN.

tower of the stables belonging to the Grands Magasins du Louvre. The stables are somewhere near the École Militaire, and would of course be very easily overlooked by the public; so to prevent this the proprietors have erected a tall slender tower, in shape somewhat resembling the clock-tower of the Houses of Parliament. It has much gilding about the top, and glitters in the sunshine like its great neighbor the dome of the Invalides. It is visible all the way from Passy to the Place de la Concorde, and from many other places besides, so that thousands of people see it every day, and many of them ask what it is. The Printemps has established its stores in a new edifice with gilded domes at the corners, in pursuance of the same policy. If the great shopkeepers found it worth while to spend money on really fine architecture, instead of scattering it about in hideous mural advertisements, the change would be most beneficial.

The magnificence of the great Parisian streets results from the habit of living in flats, as by this system a single house produces a large rental, which enables the builder to give it a magnificent front. It is obvious also that the superposition of dwellings is very favorable to height, and height is a great element of nobility in architecture. There is, however, a limit beyond which the height of houses may become injurious to the effect of a street by excluding light, and injurious also to public buildings by making them seem low. There is a tendency in London to carry houses with flats to an altitude that is desirable neither for beauty nor security. This inconvenience has been prevented in Paris by police regulations. The Prefect of Police is empowered to fix the height of houses.

It must, I fear, be admitted that the system of living on flats is likely to prevail more and more in great cities. It is, in fact, the only practical way of reconciling wide streets with a dense population. Parisians look upon it as simply rational, and they can point to their own city as evidence of the apparent spaciousness which results from it, for many of the streets and avenues are so broad that it seems as if land were of little value. The excellence of the system as regards external appearance and facility of communication is indisputable. When the population is piled high it occupies less ground and the distances are reduced. Streets that are at the same time both broad and regular in their breadth are always preferred by coachmen. On this point I have sometimes taken the opinion of Parisian drivers, and they always agreed that the new streets had immensely facilitated their work. Tramways, also, can be established in such streets; in the old narrow ones they are impossible. Broad, open spaces are favorable to public health, by giving to rooms that look out upon them as much light as if they were in the country, and almost as much air. Foot-passengers run no risk of accident except at crossings, while on narrow causeways the risks are continual. The system, then, is perfect so far as the street is concerned, and has some other great recommendations, but it is not altogether favorable to the dwelling. The dwellings are small, and the sense of

The Streets.

confinement in them is oppressive to any one who has been accustomed to space and liberty. Rents are so high that every family not positively rich is reduced to shifts and expedients. I know a young woman in the hills of the Morvan who went to Paris as a wet-nurse. She was in the family of an independent gentleman, with small or moderate means and eight children. He might have lived in the country quite at his ease, but the attraction of Paris was too great and he could not leave the capital. He had a small *appartement* at a great height, consisting of three rooms, a passage, and a little kitchen. At night all the rooms, including the passage, were converted into dormitories. The servants slept in the passage. We know what overcrowding is in London; it is a terrible evil, but it affects the poor only, while in Paris it affects the middle classes also. The evil would be still greater if the Parisians were not so excessively ingenious in the economy of space; but they are like sailors, in that they make use of every available corner. A practical result, as affecting hospitality, is that the middle-class Parisian can very rarely invite a friend to stay with him. The friend stays at some hotel, and is invited to the table only. Frequently the dining-room and kitchen are so small that it is found more convenient to dispense hospitality at a restaurant. These are real evils, but not perhaps very serious evils; the most serious evils of the system are those that affect old persons and invalids. People in weak health often remain confined in a high lodging for months together, when if they lived nearer to the ground and possessed a garden they might go into it every day. The interminable stairs have a deterrent effect on all except robust visitors, and are a real obstacle to human intercourse. Perhaps the system of superposed habitations has not yet attained its perfection. It may be that in the future there will be an extensive system of perfectly safe lifts, and it may be possible to have gardens on the roofs.

The houses are admirably lighted from the streets, and on that side they have plenty of air, but the back windows look upon narrow courts, often mere wells, which the great height of the houses makes gloomy. Once, for a fortnight, I had a room that looked into one of those wells, and the effect was so depressing that I should have preferred the poorest cottage in the country. In all other respects the new houses are a great improvement on those built just before the time of Louis Napoleon, and beyond all comparison superior to the picturesque but ill-contrived tenements of the Middle Ages.

There was one characteristic of Paris in the early part of the present century that has disappeared from the new streets: the old houses were so built, intentionally, that the fronts leaned back, sometimes with a curve that was very agreeable to artists. When Girtin went to Paris and made his sketches this inclination of the front was very common. You find it again in the etchings of Méryon and Lalanne. In contemporary street architecture it has been entirely abandoned for the perpendicular. There is another change of at least equal importance. Before Louis Napoleon the houses were

generally of unequal height, but the love of the regular
line made Haussmann's Paris almost as regular at the
cornice as at the curbstone. These changes no doubt
give a more orderly appearance to the city, but they
detract sadly from its picturesque variety. In old Paris
there were three distinct and notable irregularities:
those in the tops of the houses, the slope of the fronts,
and the ground-plan of the street, all of which are now
replaced by straight lines. In some of the new streets
the straight line is exceedingly wearisome; it is so in the
Rue de Rivoli. The reader will probably remember
the passage in Mr. Arnold's essay on "The Literary
Influence of Academies," where he criticises Mr. Palgrave for naming the feeble frivolity of the Rue de
Rivoli along with "the dead monotony of Gower or
Harley Street, or the pale commonplace of Belgrave,
Tyburnia, and Kensington." Mr. Arnold said that "the
architecture of the Rue de Rivoli expresses show, splendor, pleasure, — unworthy things, perhaps, to express
alone and for their own sakes, but it expresses them;
whereas the architecture of Gower Street and Belgravia
merely expresses the impotence of the architect to express anything." At the time when these criticisms were
written the Rue de Rivoli occupied a very different rank
among modern Parisian streets from that which it occupies at present. After the Boulevard Malesherbes, the
Boulevard Haussmann, and the Avenue Friedland, the
Rue de Rivoli, especially from the Place de la Concorde
to the Rue du Louvre, appears, I should say, rather a
street for business than anything else. The architec-

ture is decent, but plain in the extreme. There is first a simple arcade, not on pillars with pretty capitals, but on plain square stone piers. Above the arches runs a cornice that is a balcony, and carries a simple iron railing. The windows of the first floor have entablatures without sculpture, those of the second have none. On the third floor runs another balcony without ornament. I do not see either frivolity or pleasure here; it would be scarcely possible to design anything more rigid in its severity. The houses might be a line of military barracks. Eastward of the Rue du Louvre the arcade comes to an end, and the fronts of the houses become more varied. After the construction of this street the architects seem to have perceived that the mechanical repetition of the same *bay*, the same arch with the same windows above it, might ultimately be carried too far; so, happily for the future of Paris, it was thought that the Rue de Rivoli and two or three little streets close to it were a sufficient supply of identical arches with windows and cornices running to a vanishing point, like an illustration in elementary perspective.

The Avenue de l'Opéra is much finer than the Rue de Rivoli, and owes much of its superiority to the variety of its architecture. It is really a pleasure to walk quietly down one side and study the architecture over the way. As I did this once with an old French gentleman, who always foresees evil for his country, he lamented to me that the taste for material luxury should have become so predominant. To me it seems that a love for beautiful architecture is of all possible tastes the

AVENUE FRIEDLAND.

least likely to be injurious in a wealthy nation. The satisfaction it affords is purely artistic and intellectual. The carved stones are not couches of ease to lie down upon, nor dishes to pamper the appetite; they belong to the poorest as well as the richest of the citizens. All that can be reasonably objected to is the waste of wealth in the repetition of forms that have no meaning, and that are simply customary. Even incongruous innovations may sometimes be useful as an interruption to what we see every day. Somebody has built a Moorish house in the Avenue de Friedland, which, though out of place there, strikes the eye with a change that is not unwelcome. The stately, separate mansions are a great relief after the continuous blocks of buildings.

It is much to be regretted that many fine old houses have disappeared, but a few are visible still. I remember the feeling of sudden and keen pleasure with which I first came upon the Hôtel La Valette on the Quai des Célestins. The restoration of it had been begun with the intention of making it a private museum, but it has changed hands and is now a school. These old houses are seldom preserved as residences, and the best that can happen to them is to be employed as museums, like the Hôtel Carnavalet, which is to be the future lapidary museum of Paris and library of historical records concerning the history of the city.[1] This project was due to Baron Haussmann, the great destroyer of

[1] A description of the Hôtel Carnavalet, with an account of the museum, was published in "L'Art" for January 18 and January 25, 1880. The hotel is situated at the angles of the Rue des Francs Bourgeois and the Rue Sevigné, not far from the Rue Turenne.

old Paris, who in this instance appears as a preserver. The architecture of the hotel is heavy, but would appear much heavier if it were not lightened by the graceful sculpture of Jean Goujon. One of the curiosi-

HÔTEL DE SENS.

ties of Paris in domestic architecture is the house on the Cours la Reine, called the *Maison de François I.* Here we have an excellent example of what ought to have been done with many old houses. This one was erected near Fontainebleau by Francis I., and sold in 1826 to a private purchaser, who had every stone removed to Paris and erected again, as we see. The house is not large, but the size of it is practically much increased by its having been rebuilt on a broad basement that gives a terrace round the building without injuring its architecture in the slightest degree, while it affords ample room for kitchens and other offices and leaves the beautiful little house itself for the master and his family. In the front are three arches with a broad frieze above them. Above the frieze are three windows, very large in proportion, as they are divided only by piers the width of the pilasters in front of them and by their own heavy mullions and transoms. Over the windows is an entablature, and the whole is crowned with a parapet which is pierced over the windows, but not elsewhere, a refinement clearly demonstrating the artist-nature of the architect. There is no visible roof, but the need for one is not felt. The front is rich in beautiful sculpture, supposed to be by Jean Goujon, and including medallion-portraits of royal personages. There are also decorative trophies and subjects illustrating the vintage.

Although the Hôtel de Cluny has not been transferred to another site like the Maison de François I., it has been almost as wonderfully preserved. It was built at first by the Abbots of Cluny, but not much used by them. In the early part of the present century it was private property let in tenements to a number of tenants. It now belongs to the State, a happy result due entirely to the public spirit of a lady, Madame du Sommerard, widow of the antiquary and collector who had found a home for his treasures in the Hôtel de Cluny, which he owned. Madame du Sommerard sold the whole together to the State at a loss to herself, as she had much more advantageous offers. Thus it has most happily come to pass that in the midst of a very busy part of Paris, close to the great Boulevards of St. Germain and St. Michel, there is a safe little island of the past amid the noisy torrents of the present. I know nothing more delightful in Paris than the peace of the Hôtel de Cluny; and what a wonderful piece of good luck it is that this beautiful relic of the fifteenth century should have been quite close to the most interesting remnant of Roman Paris, so that both can be kept together in the same safe enclosure! I have only space to point to a few of the chief characteristics of the building. I do not know of any kind of domestic architecture quite so satisfactory as that when the house is isolated. For street architecture the modern Parisian is practically much better; but for a builder who has but one dwelling to erect, and is not restricted to ground-space, this fifteenth-century architecture is the one that best unites a homely expression with beauty and convenience. The walls are not too high, the roof has a comfortable appearance, the building is of ample size yet not wearisome in vastness; it is not a proud palace, but a beautiful home that one might live in habitually and love with intense affection.

The Streets. 233

The windows in the walls are square-headed, with mullions, transoms, and weather-mouldings that connect the windows together. There is a pierced parapet, and the dormer-windows are beautifully finished with pinnacles and finials. There are several staircase turrets. It is beyond my province here to speak of collections, but those in the Hôtel de Cluny, illustrating the Middle Ages and the Renaissance, are as interesting as the delightful building that contains them. The Louvre is the place to study sculpture, but the lover of *carving* (in stone, wood, and ivory) should go to the Hôtel de Cluny. The other beautiful example of fifteenth-century domestic architecture, the Hôtel de Sens, also built by a great ecclesiastic (the Archbishop of Sens) for his town residence, is remarkable for the great development of bartizan turrets relatively to the rest of the building. I do not know of any edifice whatever in which they are relatively so large; but as they are enriched with panels and carving, the size of them may be forgiven. They have become very familiar objects of late years, as the hotel is unfortunately occupied as a manufactory of sweets, and the enterprising maker uses a representation of the building in all his illustrated advertisements. How little the architect in the fifteenth century foresaw this special kind of celebrity for his work!

There is a very curious example of the modern love of symmetry and order in the arrangements concerning the outside of St. Germain l'Auxerrois. As you stand under the eastern entrance to the old courtyard of the Louvre, the front of that church is to your right on the other side the Rue du Louvre, but it is not parallel with the street; it inclines towards the east. It was a very perplexing problem to get any symmetry out of that, but the solution was found in the construction of another building — a Mairie — inclined conversely, and in the erection of a tower between the two. St. Germain l'Auxerrois is in Gothic, and the Mairie is in a modified Renaissance; yet the architect has had the art and skill to give, in Renaissance, an echo of the Gothic ideas in the church, so that there is a strong general resemblance between the Mairie and the church, in spite of the difference of style.

The vast increase of wealth and luxury in Paris during the present century has led to the construction of a great number of isolated dwellings, many of which would deserve study as examples of modern house-architecture. They are generally much superior to the London villa in elegance of design and in the quality and genuineness of the materials employed. The best of them are to be found in the regions near the Bois de Boulogne, especially about Passy and Auteuil. There are some particularly good examples on the Boulevard Beauséjour and the Boulevard de Montmorency. The misfortune of most residences of that kind is that they are almost sure to be injured by the too near neighborhood of others. I remember a house on the Boulevard Beauséjour which is of classic design and in very perfect taste, but it happens to be low and close to a lofty edifice that crushes it completely. Again, from the variety of styles adopted, it may easily happen that you cannot attune

THE MAIRIE AND ST. GERMAIN L'AUXERROIS.

your mind to the enjoyment of one style because a style with opposite qualities is forcing itself upon your attention at the same time. Formerly, when land was cheaper, there were many isolated houses within the fortifications which stood in their own little parks, quite separated from others by groves of shady trees. These little parks are becoming fewer every day. Where one villa stood thirty years ago three stand now, and sometimes half-a-dozen. Besides this, the old region for villas — Auteuil — is becoming a town like Passy. Enormous blocks of new houses, as large and lofty as any in the heart of Paris, are rising on the park lands and cutting them into formal streets. An old friend of mine had a delicious retreat at Auteuil, — a small house in a large space of grass and grove. I went to find it this year, and found a block of buildings six stories high and as long as the Hôtel du Louvre.

The tendency of the French towards orderly and methodical arrangement is exemplified nowhere more strongly than in the radiation of the avenues from the Arc de l'Étoile. That huge triumphal arch is admirably situated on its height, and the ediles appear to have determined that it should be seen from as many points as possible. There is no more stately arrangement in any capital than the wheel of streets that radiate from that wonderful centre. There are twelve of them, three of which are a hundred mètres wide, while seven of them are more than a thousand mètres long, and in five directions there is a clear view of more than an English mile. Such sort of beauty and sublimity as the straight and broad street has to offer, with its interminable rows of trees, its vast causeways, its lofty houses, has surely been here attained, if anywhere. I admit the grandeur, the masterful thoroughness with which the idea has been carried out, but never felt the slightest desire to live in streets so totally wanting in homeliness. Many a snug, unpretending old house in some dull provincial town has inspired me with a sudden, almost envious affection; but in these wearisome long avenues the best thing seems to be the tram-car that carries one well to the end of them.

A question very nearly affecting the appearance of Parisian streets is at this date (1885) looming in the immediate future. Paris is to have internal railways. Commissioners have examined our London underground system, and they have also examined the American aerial system. For a long time the decision was uncertain, although it was confidently announced that the American system had been adopted. Now, however, it appears that the three powers, the Government, the Municipal Council of Paris, and the *Conseil Général des Ponts et Chaussées* are finally of one mind upon the subject, and before these words are printed it is likely that their scheme will have received the assent of the Chambers. It includes a great line traversing Paris from east to west on the right bank of the Seine, and also a great line crossing this from north to south. Besides these, there will be a curved line on the left bank, and the plan leaves room for extensions. The important question whether the new metropolitan railway is to be sub-

RUE DES CHIFFONNIERS, PARIS. DRAWN BY LEON LHERMITTE.

terranean as in London, or above the streets as in New York, is decided in favor of a subterranean line for all the more crowded parts of the city, which leaves a latitude for the aerial system elsewhere. Every one who cares for the beauty of the most beautiful modern capital must learn with apprehension that aerial railways will be tolerated in it anywhere. Certainly there is a degree of architectural taste and knowledge in Paris that may preserve it from the engineering monstrosities which England and America tolerate, and it is probable that if aerial railways are made at all, they will be designed with as much art and care as the nature of the erection will permit. Still, a row of cast-iron columns supporting an endless bridge in two stories can hardly be otherwise than mechanically monotonous.[1]

Here may end this series of chapters on Paris, with regard to which the writer is clearly aware that so vast a subject cannot be treated without many omissions.[2] He has principally concerned himself with its artistic aspects, and has only made occasional reference to the

[1] So far as the scheme is known hitherto, the aerial railway is to be a double line in which one pair of rails will be placed above the other.

[2] For example, I have omitted the Palais Royal, but that is chiefly interesting historically; the present building is of little architectural importance, and the little shops in the square that were such an attraction in the time of our grandfathers are eclipsed by others in the new streets. I should have liked to mention some fountains and other things, but it is most difficult to compress accurate description, and criticism that gives reasons, within narrow limits. The Church of St. Germain des Prés would have been an excellent subject for a chapter if the old Romanesque edifice had been preserved in its integrity, but as that is not the case I preferred to speak of churches fully representative of their styles.

far wider historical and social aspects, concerning which the reader may find abundance of information elsewhere. Paris, as it exists at present, is the model modern city that others copy, and that London herself is probably destined to copy when the density of population makes it more and more necessary to pile many human beings on a square mile, without impeding a constantly increasing circulation. What is chiefly to be regretted in the French capital is, that of the beautiful mediaeval city that preceded it so little — and that only in isolated specimens — has been preserved to our own time. The Present is merciless to the Past, and merciless it has always been. It may, however, be truly said that our age shows less of this mercilessness than its predecessors. When they preserved things it was chiefly from carelessness and indolence; but we preserve, when we think of it to preserve at all, from artistic or archaeological interest. No century but our own ever made intentional sacrifices for the preservation of ancient monuments. The nineteenth century has made some sacrifices of this kind; its shame is that they have been so few.

www.ingramcontent.com/pod-product-compliance
Lightning Source LLC
Chambersburg PA
CBHW030314240426
43673CB00040B/1156